Published for
**OXFORD INTERNATIONAL
AQA EXAMINATIONS**

International GCSE
CHEMISTRY
Revision Guide

Adam Robbins
Adam Boxer
Philippa Gardom-Hulme
Editor: Primrose Kitten
Elizabeth McCullough

OXFORD
UNIVERSITY PRESS

Contents

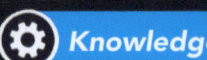  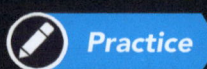 

 Shade in each level of the circle as you feel more confident and ready for your exam.

How to use this book iv

| C1 Atomic structure | 2 |
| Knowledge |
| Retrieval |
| Practice |

| C2 Structure and bonding | 14 |
| Knowledge |
| Retrieval |
| Practice |

| C3 The Periodic Table | 26 |
| Knowledge |
| Retrieval |
| Practice |

| C4 Metals | 38 |
| Knowledge |
| Retrieval |
| Practice |

| C5 Electrolysis | 50 |
| Knowledge |
| Retrieval |
| Practice |

| C6 Chemical analysis | 62 |
| Knowledge |
| Retrieval |
| Practice |

| C7 Acids, bases, and salts | 74 |
| Knowledge |
| Retrieval |
| Practice |

| C8 Quantitative chemistry | 84 |
| Knowledge |
| Retrieval |
| Practice |

C9 Rates of reactions 100	**C12 Carbon compounds as fuels** 136
Knowledge ⊖	Knowledge ⊖
Retrieval ⊖	Retrieval ⊖
Practice ⊖	Practice ⊖
C10 The extent of reactions 112	**C13 Other hydrocarbon products** 148
Knowledge ⊖	Knowledge ⊖
Retrieval ⊖	Retrieval ⊖
Practice ⊖	Practice ⊖
C11 Energy changes 124	**C14 Alcohols, carboxylic acids, and esters** 158
Knowledge ⊖	Knowledge ⊖
Retrieval ⊖	Retrieval ⊖
Practice ⊖	Practice ⊖

Periodic Table 172

Answers

All of the **answers** are on the website at www.oxfordsecondary.com/oxfordaqa-revision

How to use this book

This book uses a three-step approach to revision: **Knowledge**, **Retrieval**, and **Practice**.
It is important that you do all three; they work together to make your revision effective.

1 Knowledge

Knowledge comes first. Each chapter starts with a **Knowledge Organiser**. These are clear, easy-to-understand, concise summaries of the content that you need to know for your exam. The information is organised to show how one idea flows into the next so you can learn how all the science is tied together, rather than lots of disconnected facts.

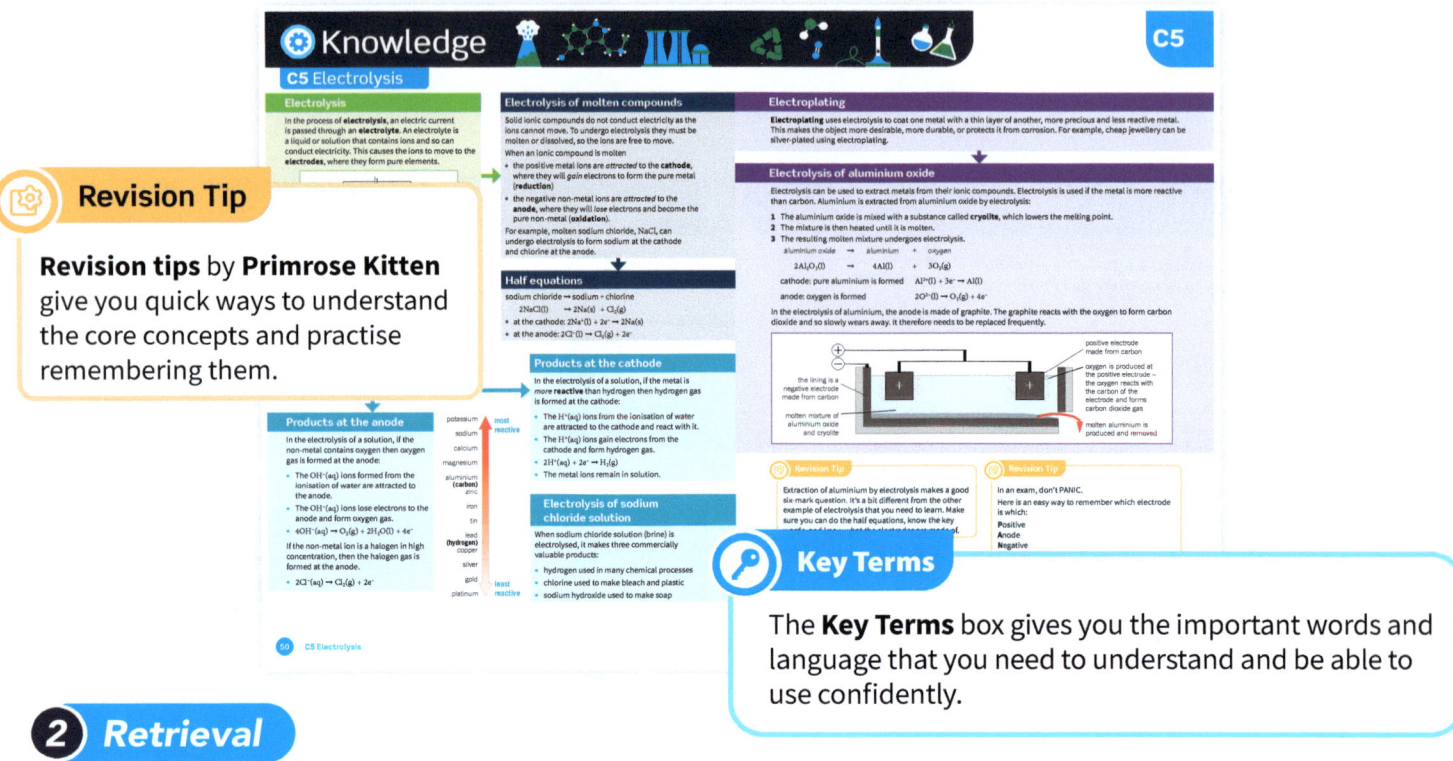

Revision Tip

Revision tips by **Primrose Kitten** give you quick ways to understand the core concepts and practise remembering them.

Key Terms

The **Key Terms** box gives you the important words and language that you need to understand and be able to use confidently.

2 Retrieval

The **Retrieval questions** help you learn and quickly recall the information you've acquired. These are short questions and answers about the content in the Knowledge Organiser. Cover up the answers with some paper and write down as many answers as you can from memory. Check back to the Knowledge Organiser for any you got wrong, then cover the answers and attempt *all* the questions again until you can answer all the questions correctly.

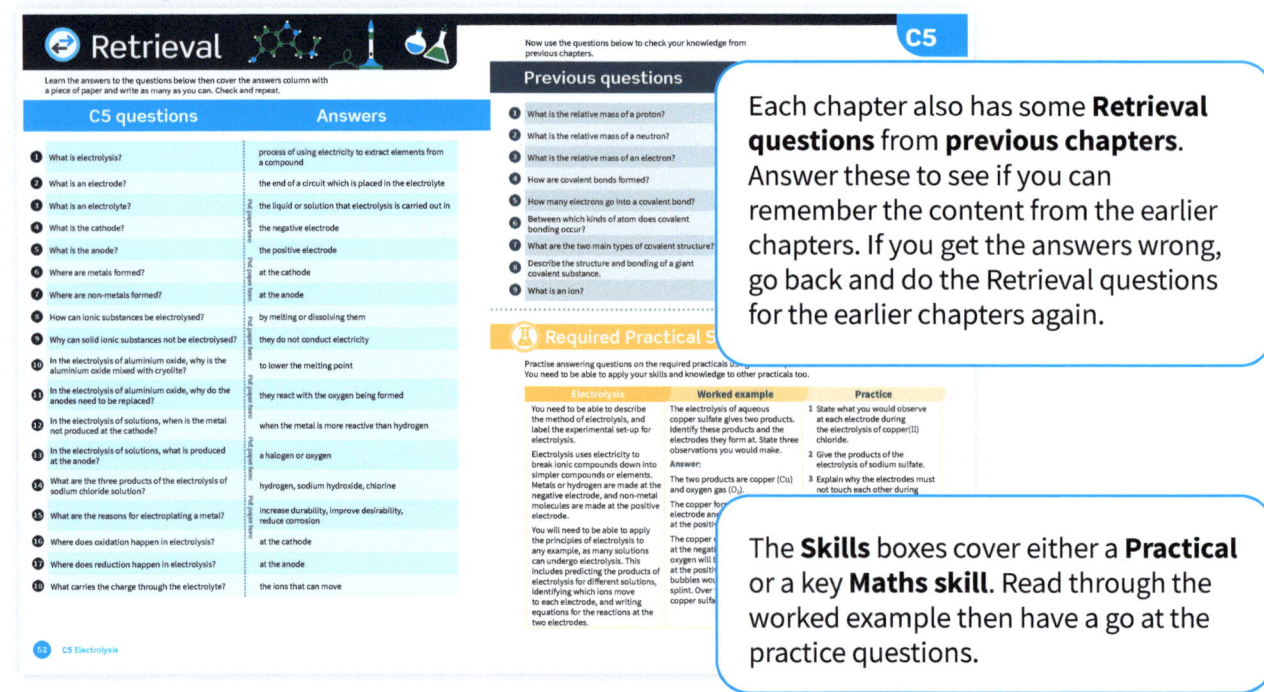

Each chapter also has some **Retrieval questions** from **previous chapters**. Answer these to see if you can remember the content from the earlier chapters. If you get the answers wrong, go back and do the Retrieval questions for the earlier chapters again.

The **Skills** boxes cover either a **Practical** or a key **Maths skill**. Read through the worked example then have a go at the practice questions.

Make sure you revisit the retrieval questions on different days to help them stick in your memory. You need to write down the answers each time, or say them out loud, otherwise it won't work.

3 Practice

Once you think you know the Knowledge Organiser and Retrieval answers really well you can move on to the final stage: **Practice**.

Each chapter has lots of **exam-style questions** to help you apply all the knowledge you have learnt and can retrieve.

Each question has a difficulty icon that shows the level of challenge.

 These questions build your confidence.

 These questions consolidate your knowledge.

 These questions stretch your understanding.

Make sure you attempt all of the questions no matter what grade you are aiming for.

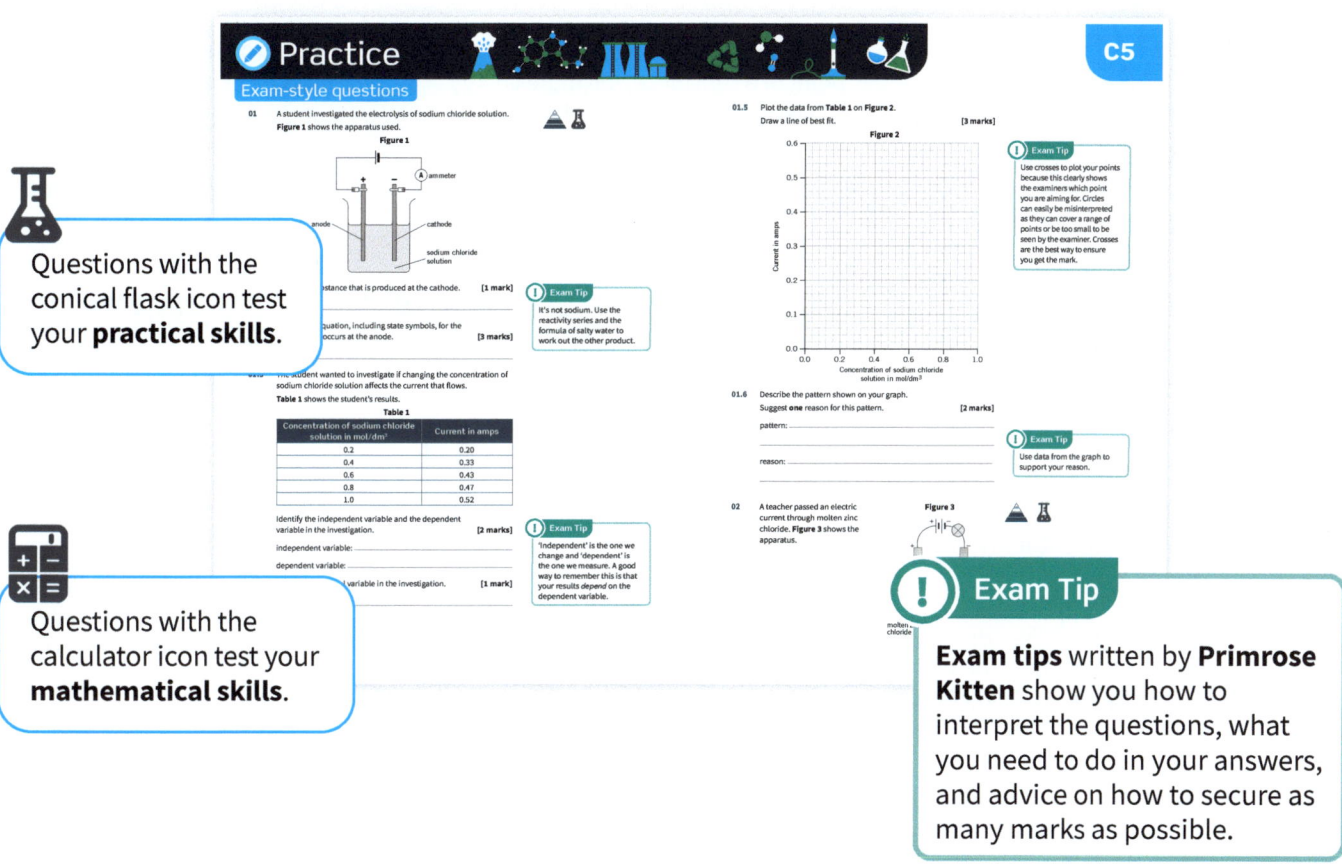

Questions with the conical flask icon test your **practical skills**.

Questions with the calculator icon test your **mathematical skills**.

Exam tips written by **Primrose Kitten** show you how to interpret the questions, what you need to do in your answers, and advice on how to secure as many marks as possible.

Knowledge

C1 Atomic structure

Particle model

The three states of matter can be represented in the particle model.

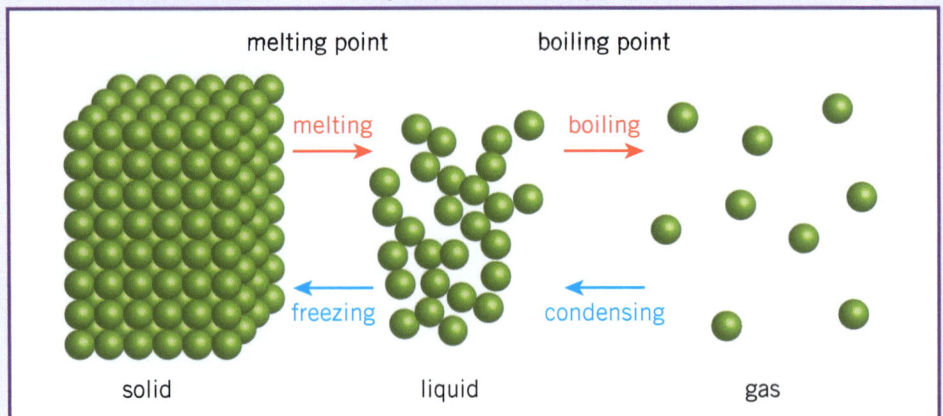

This model assumes that:
- there are no forces between the particles
- that all particles in a substance are spherical
- that the spheres are solid.

The amount of energy needed to change the state of a substance depends on the forces between the particles. The stronger the forces between the particles, the higher the melting or boiling point of the substance.

> **Revision Tip**
>
> The higher the melting or boiling point of a substance, the more energy is needed for the change of state.

Atoms, elements, and compounds

All substances are made of **atoms**.

Elements are substances made of only one type of atom. Each atom of an element will have the same number of protons in the **nucleus**. Elements are shown in the **Periodic Table**. Atoms of each element are represented by a chemical symbol, e.g., O represents an atom of oxygen.

Compounds are made of two or more different types of atoms chemically bonded together. The atoms in a compound have different numbers of protons.

Sub-atomic discoveries

The discovery of electrons allowed scientists to work out that elements with the same number of electrons in their outer shell had similar **chemical properties**.

The discovery of protons allowed scientists to order the elements in the Periodic Table by their atomic number.

The discovery of neutrons led to scientists discovering isotopes. Isotopes explained why some elements didn't seem to fit when the Periodic Table was organised by atomic mass (like iodine and tellurium).

Atomic and mass numbers

The Perodic Table shows the atomic number and mass number of each element.

Atomic number is the number of protons in an atom of that element. Mass number is the total number of protons and neutrons in an atom of that element.

7 — mass number
Li
3 — atomic number

Relative charge and mass

	Relative charge	Relative mass	
Proton	+1	1	= **atomic number**
Neutron	0	1	= mass number − atomic number
Electron	−1	0 (very small)	= same as the number of protons

All atoms have equal numbers of protons and electrons, meaning they have no overall charge:

total negative charge from electrons = total positive charge from protons

C1

The structure of the atom

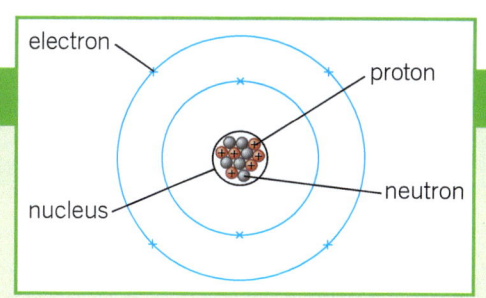

Atoms have a small dense nucleus made of protons and neutrons. They then have electrons orbiting on energy levels (also called shells). The attraction between the protons in the nucleus and the electrons prevents them from escaping.

Drawing atoms

Electrons in an atom are placed in fixed **shells**, or **energy levels**. We represent electrons using dots or crosses. You can put:
- up to two electrons in the first shell
- eight electrons each in the second and third shells.

lithium   chlorine

You must fill up a shell before moving on to the next one. You can use a shorthand to show the electron configuration. Write the number of electrons in each shell separated by a comma, starting with the first shell. For example, Li is 2,1 and Cl is 2,8,7.

Relative atomic mass

All relative atomic masses are relative to the mass of an atom of ^{12}C, which has a mass of exactly 12.

relative atomic mass,

$$A_r = \frac{\text{average mass of 1 atom}}{\frac{1}{12} \text{ mass of 1 atom of } ^{12}C}$$

Isotopes

Atoms of the same element can have a different number of neutrons, giving them a different overall mass number. Atoms of the same element with different numbers of neutrons are called **isotopes**.

The **relative atomic mass** is the average mass of all the atoms of an element (note that **abundance** means the percentage of atoms with a certain mass):

$$\text{relative atomic mass} = \frac{(\text{abundance of isotope 1} \times \text{mass of isotope 1}) + (\text{abundance of isotope 2} \times \text{mass of isotope 2})...}{100}$$

Diffusion

In liquids and gases, the random movement of particles mixes substances in a process called diffusion.

Diffusion takes place faster in a gas than in a liquid. Small, light particles diffuse faster than large, heavy ones.

Examples are:
1. potassium permanganate(VII) in water
2. ammonia and hydrochloric acid
3. bromine in air.

Key Terms

Make sure you can write a definition for these key terms.

abundance   atom   atomic number   compound   electron   element   energy level   isotope   neutron
nucleus   proton   relative atomic mass   relative charge   relative mass   shell

C1 Knowledge   3

Retrieval

Learn the answers to the questions below then cover the answers column with a piece of paper and write down as many as you can. Check and repeat.

	C1 questions	Answers
1	What is an atom?	smallest part of an element that can exist
2	What is an element?	substance made of one type of atom
3	What do we call the change of state when we heat a liquid?	boiling
4	What do we call the change of state when a gas cools?	condensation
5	Where are protons and neutrons?	in the nucleus
6	What is the relative mass of a proton?	1
7	What is the relative mass of a neutron?	1
8	What is the relative mass of an electron?	0 (very small)
9	What is the relative charge of a proton?	+1
10	What is the relative charge of a neutron?	0
11	What is the relative charge of an electron?	−1
12	How can you find out the number of protons in an atom?	the atomic number on the Periodic Table
13	How can you calculate the number of neutrons in an atom?	mass number − atomic number
14	Why do atoms have no overall charge?	equal numbers of positive protons and negative electrons
15	How many electrons would you place in the first, second, and third shells?	up to 2 in the first shell and up to 8 in the second and third shells
16	What is a compound?	substance made of more than one type of atom chemically joined together
17	Where are all the elements recorded?	on the Periodic Table
18	What are isotopes?	atoms of the same element (same number of protons) with different numbers of neutrons
19	Describe the structure of the atom.	a dense nucleus of protons and neutrons with electrons orbiting around
20	What is relative mass?	the average mass of all the atoms of an element

C1 Atomic structure

Maths Skills

Practise your maths skills using the worked example and practice questions below.

Unit conversion

Scientists use different units depending on what is most useful to them. For example, when talking about the size of molecules it doesn't make sense to talk about them in kilometres, so we can use nanometres instead.

Whenever we do a calculation, we need to make sure the units are the same, so have to do a unit conversion.

The table below shows you how some units can be compared to each other.

Unit	Standard form in m
1 metre (m)	1×10^{0} m
1 centimetre (cm)	1×10^{-2} m
1 millimetre (mm)	1×10^{-3} m
1 micrometre (μm)	1×10^{-6} m
1 nanometre (nm)	1×10^{-9} m
1 picometre (pm)	1×10^{-12} m

Worked example

Express 120 cm in metres.

When converting to a larger unit, multiply the original value by the value in metres in standard form.

$120 \times 1\times10^{-2} = 1.2$ m

Express 120 m in centimetres.

When converting to a smaller unit, divide the original value by the value in metres in standard form.

$= \dfrac{120}{1\times10^{-2}} = 12\,000$ cm

Practice

1. Express 400 cm in metres.
2. Express 20 m in millimetres.
3. Express 0.8 m in nanometres.

Practice

Exam-style questions

01 **Table 1** gives some information about four different atoms. The atoms are represented by the letters **W**, **X**, **Y**, and **Z**. These letters are not the chemical symbols of the elements.

Table 1

Atom	Number of protons	Number of neutrons	Number of electrons
W	16	16	
X	17	20	17
Y	18	22	18
Z	17	18	17

01.1 Give the number of electrons in atom **W**. [1 mark]

01.2 Give the atomic number of atom **X**. [1 mark]

> **Exam Tip**
>
> The clue in the question is that these are atoms. Use the Periodic Table to help you find the answer.

01.3 Give the letter of the atom that has the greatest mass number. [1 mark]

01.4 Give the letters of the **two** atoms that are isotopes of the same element. [1 mark]

02 Atoms are made up of sub-atomic particles.

02.1 Give the relative charge of each sub-atomic particle. [3 marks]

02.2 Identify which sub-atomic particle determines the identity of an element. [1 mark]

02.3 Using the Periodic Table, give the atomic number of oxygen. [1 mark]

6 C1 Atomic structure

C1

02.4 Explain why the mass number of chlorine is not a whole number. **[2 marks]**

Exam Tip
Chlorine has two common isotopes.

02.5 Write the electronic structure of a phosphorus atom. **[1 mark]**

03 Phosphorus has 15 electrons.

03.1 Sketch the electronic structure of phosphorus. **[1 mark]**

Exam Tip
This is science not art, so your circles don't need to be perfect!

03.2 Deduce the group number in the Periodic Table that phosphorus is in. **[1 mark]**

03.3 Compare the properties of protons, neutrons, and electrons. Include in your answer the location of each type of sub-atomic particle within an atom. **[6 marks]**

Exam Tip
There are 6 marks available for this answer and the question is asking about three different sub-atomic particles. So you need write a balanced answer covering all three equally.

04 Potassium chloride is a salt. A student has a sample of potassium chloride solution.

04.1 Potassium chloride is a solid. Complete the particle diagram in **Figure 1** to represent solid potassium chloride. Use a circle to represent each potassium chloride particle. **[3 marks]**

Figure 1

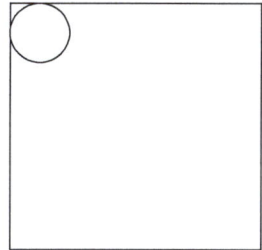

04.2 Describe how you can tell that potassium chloride is a compound and not a mixture of potassium and chlorine. **[1 mark]**

05 **Figure 2** shows the electronic structure of an atom. The atom has no overall charge.

Figure 2

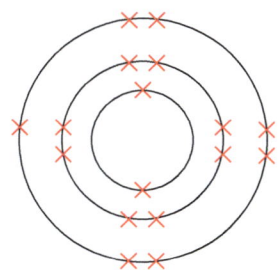

C1 Practice 7

05.1 Identify the number of protons in the nucleus of the atom. **[1 mark]**

05.2 Give the atomic number of the atom. **[1 mark]**

05.3 **Figure 2** shows a chlorine atom. Chlorine has a relative atomic mass of 35.5. Explain why the relative atomic mass of chlorine is not a whole number. **[2 marks]**

> **Exam Tip**
>
> Remember protons have a positive charge and electrons have a negative charge.

06 An atom of silicon has 14 electrons.

06.1 Give the relative charge of an electron. **[1 mark]**

06.2 **Figure 3** shows the energy levels (shells) of the electrons in a silicon atom. Complete the diagram by drawing the 14 electrons in the correct shells. **[1 mark]**

Figure 3

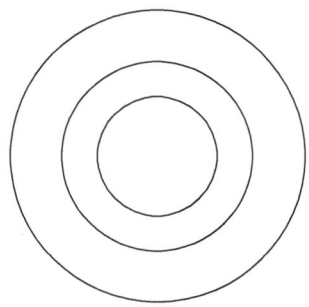

> **Exam Tip**
>
> Start from the centre and work your way out.

06.3 Silicon can exist as different isotopes. Explain why all isotopes of silicon have the same chemical properties. **[1 mark]**

07 **Table 2** gives some information about the most common isotopes of some elements.

Table 2

Element	Number of protons	Number of neutrons
neon	10	10
calcium	20	20
zinc	30	34
zirconium	40	50
tin	50	70
lanthanum	57	82

07.1 Write the name of the element that has an atomic number of 40. **[1 mark]**

07.2 Write the name of the element that has a mass number of 40. **[1 mark]**

07.3 Write the electronic structure of a calcium atom. **[1 mark]**

> **Exam Tip**
>
> Starting from the inside shell, write down the number of electrons in each shell followed by a comma: x,y,z.

07.4 Plot the data from **Table 2** as a scatter graph on **Figure 4**. **[3 marks]**

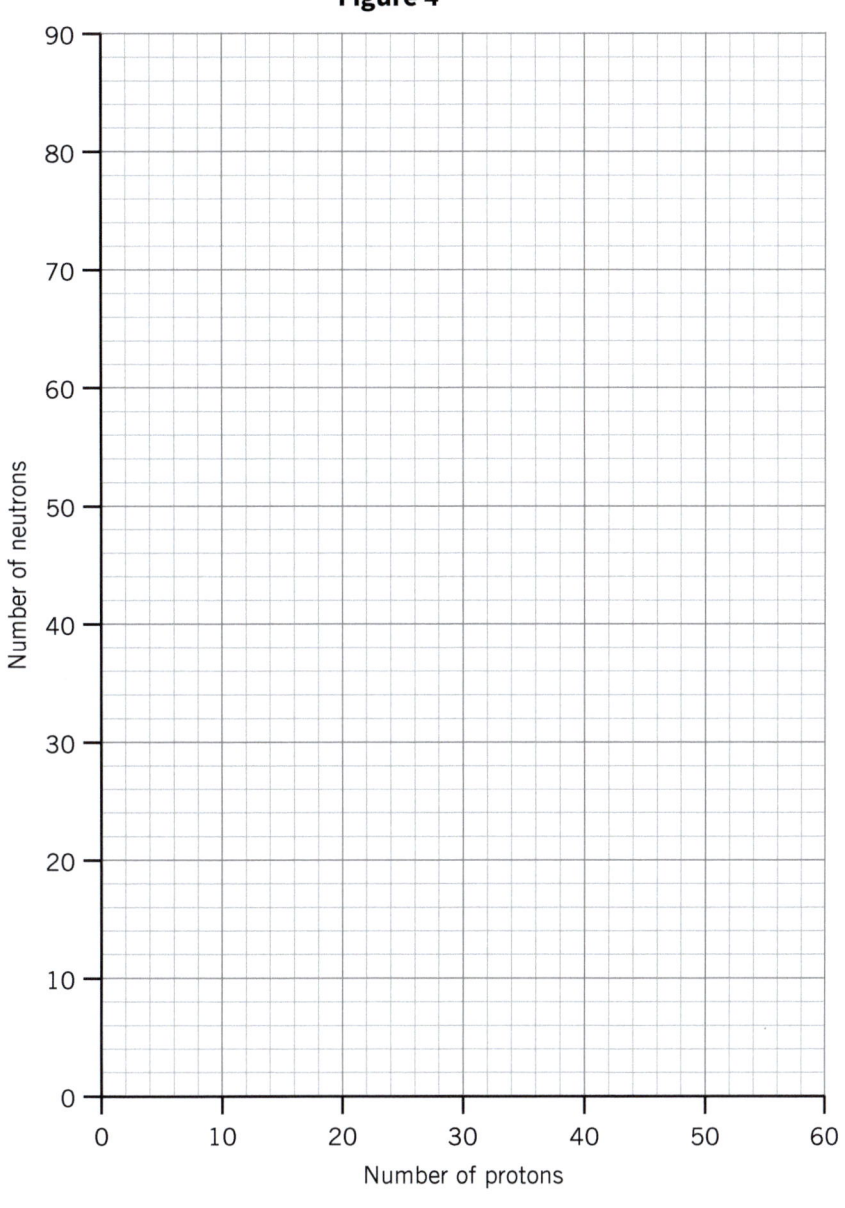

Figure 4

> **Exam Tip**
> Use crosses to show where you have plotted you points.

07.5 Draw a curve of best fit on your graph. **[1 mark]**

07.6 Describe the relationship shown on the graph. **[2 marks]**

> **Exam Tip**
> Lines of best fit need to be smooth and continuous.

08 A scientist uses the following symbols to represent some substances.
- potassium atom: ◯
- sodium atom: ◇
- chlorine atom: ◻
- water: △

08.1 Identify the symbol that represents a compound. **[1 mark]**

08.2 The scientist uses the symbols to draw a representation of four samples they have (**Figure 5**).

Figure 5

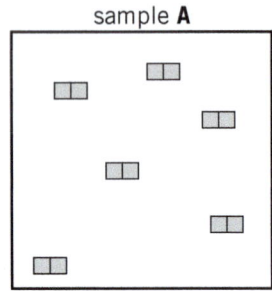
sample **A**

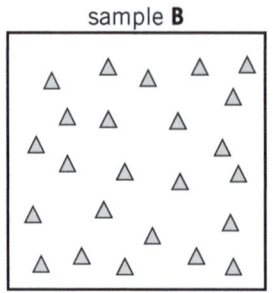
sample **B**

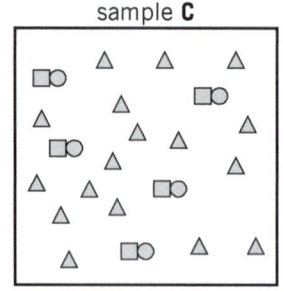
sample **C**

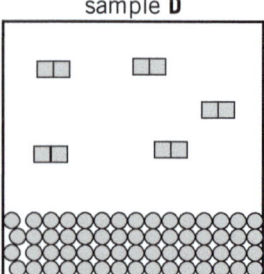
sample **D**

Identify the sample that contains a pure element. **[1 mark]**

08.3 Identify the sample that contains a mixture of two compounds. **[1 mark]**

> **Exam Tip**
> Look carefully at the key when answering these questions.

08.4 Identify the sample that contains a mixture of two elements. **[1 mark]**

08.5 The scientists uses their symbols to draw the following representation of the compound sodium chloride.

Write the chemical formula for this substance. **[1 mark]**

09 **Table 3** gives the numbers of protons, neutrons, and electrons for some atoms and ions. The atoms and ions are represented by the letters **A** to **E**.

These are not their chemical symbols. You will need to refer to the Periodic Table.

Table 3

Atom, isotope, or ion	Number of protons	Number of neutrons	Number of electrons
A	7	7	7
B	11	12	10
C	12	13	12
D	12	12	10
E	7	8	7

09.1 Write the chemical symbol of **A**, including its mass number, atomic number, and any charge. **[1 mark]**

> **Exam Tip**
> The atomic number is the clue to the chemical symbol.

09.2 Give the letter of the isotope of **A** that is shown in **Table 3**.
Write its chemical symbol, its mass number, its atomic number, and any charge. **[2 marks]**

09.3 Give the letter (**A–E**) of the ion of **C** shown in **Table 3**.
Write the chemical formula of the ion of **C**, including its charge. **[2 marks]**

10 C1 Atomic structure

09.4 Give the chemical symbol of the atom in **Table 3** that has the greatest mass number. Write its mass number and atomic number.
[2 marks]

10 **Figure 6** shows the boiling points of three substances; ethanol (C₂H₅OH), hexanol (C₆H₁₃OH), and mercury. Each substance is represented by a letter. The letters are not the chemical symbols of the substances.

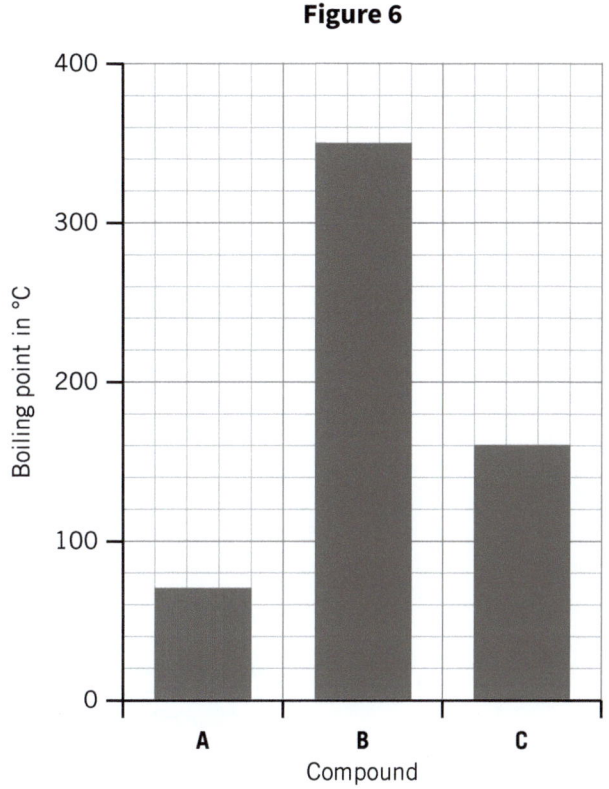

Figure 6

10.1 Give the boiling point of substance **B**. **[1 mark]**

10.2 Identify which substance has a boiling point of 78 °C. **[1 mark]**

10.3 Suggest which letter represents each substance. Explain your answer. **[3 marks]**

Exam Tip
Think about the structures of each of the compounds and how they will affect their boiling points.

11 Copper has two stable isotopes. The chemical symbols of these isotopes are $^{63}_{29}Cu$ and $^{65}_{29}Cu$.

11.1 Give the number of protons in an atom of $^{65}_{29}Cu$. **[1 mark]**

11.2 Give the number of neutrons in an atom of $^{65}_{29}Cu$. **[1 mark]**

11.3 Give the mass number of the $^{63}_{29}Cu$ atom. **[1 mark]**

12.1 Draw **one** line from each sub-atomic particle to the relative charge. **[2 marks]**

Sub-atomic particle	Relative charge
neutron	+1
proton	0
electron	−1

12.2 A particle has four protons and four electrons. Give the charge on the particle. **[1 mark]**

12.3 A particle has five protons and four electrons. Determine whether the particle is an atom or an ion. **[1 mark]**

An ion is an atom that has lost or gained electrons.

12.4 An atom has six protons in its nucleus. Complete the dot and cross diagram in **Figure 7** to show the number of electrons in the atom. **[2 marks]**

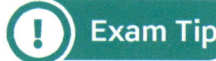

Start from the innermost shell.

Figure 7

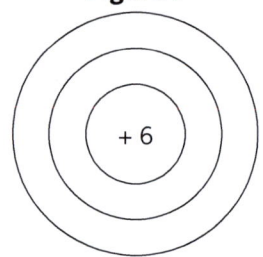

12.5 Determine the identity of the atom in **12.4**. **[1 mark]**

12.6 Complete the following sentence.

Isotopes are atoms of an element that have the same number of _____ but different number of _____.

Remember it is the atomic number – the number of protons – that determines the identity of an atom.

12.7 The chemical symbol of silicon is $^{28}_{14}Si$.

Describe an atom of silicon in terms of the number of protons, neutrons, and electrons, and their charges. **[2 marks]**

13.1 Carbon dioxide is made up of one carbon atom and two oxygen atoms. Identify the correct chemical formula for carbon dioxide. Choose **one** answer. **[1 mark]**

Co_2 cO^2 CO^2 CO_2

13.2 Identify the three elements in H_2SO_4. Use the Periodic Table to help you. **[3 marks]**

13.3 How many atoms are in one molecule of NaOH? Choose **one** answer. **[1 mark]**

1   2   3   4

Look for the capital letters! Capital letters are very important in chemical formulae. A capital letter is the difference between carbon dioxide gas and a lump of metal cobalt.

C1 Atomic structure

14 A student has the compound CuXO₄. The student does not know the identity of the element X. The relative formula mass of the compound is 159.5.

14.1 Calculate the relative atomic mass of X. Relative atomic masses A_r: Cu = 63.5   O = 16   **[3 marks]**

> **Exam Tip**
> There are four oxygens so you need to include 4 × 16 =

14.2 Use your answer to **14.1** to identify X. **[1 mark]**

14.3 The atomic number of oxygen is 8. Calculate how many neutrons are in a nucleus of oxygen. **[1 mark]**

14.4 Explain why the relative atomic mass of copper, Cu, is not a whole number. **[2 marks]**

15.1 Complete the sentences. Choose the anwers from the box. **[3 marks]**

| innermost | highest | lowest | outermost | 2 | 8 |

Electrons fill the _____ energy level first. The lowest energy levels are the _____ shells. The shell closest to the nucleus can hold up to _____ electrons.

15.2 An atom has the electronic structure 2,8,1. Is this atom boron or sodium? Use the Periodic Table to help you. **[1 mark]**

15.3 Oxygen has eight electrons. Complete the dot and cross diagram in **Figure 8** to show the electronic structure of oxygen. **[2 marks]**

> **Exam Tip**
> Start from the innermost shell and then count until you get eight electrons. Think about how many electrons each shell can hold.

Figure 8

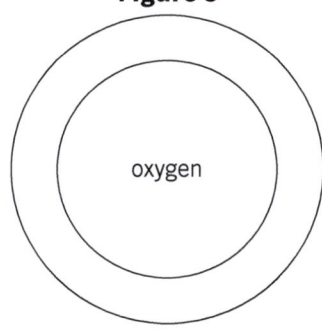

Knowledge

C2 Structure and bonding

Covalent bonding

Atoms can share or transfer electrons to form strong chemical bonds. A **covalent bond** is when electrons are *shared* between **non-metal** atoms. The number of electrons shared depends on how many extra electrons an atom needs to make a full outer shell and have the electron arrangement of a noble gas. If you include electrons that are shared between atoms, each atom has a full outer shell.

Single bond = each atom shares one pair of electrons.
Double bond = each atom shares two pairs of electrons.

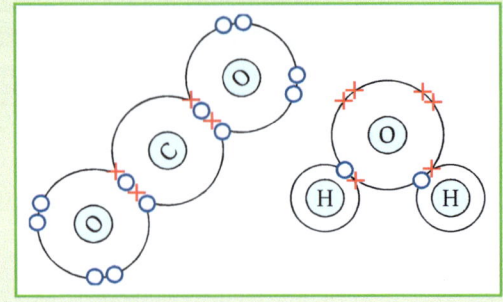

Covalent structures

When atoms form covalent bonds, different types of structures can be formed. The structure depends on how many atoms there are and how they are bonded. There are two main types of covalent structure:

	Giant covalent	Small molecules
Structure and bonding	Many billions of atoms, each one with a strong covalent bond to a number of others. For example, diamond is a giant covalent structure. 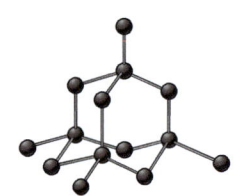	Each molecule contains a few atoms with strong covalent bonds between them. Different molecules are held together by weak **intermolecular forces**. For example, water is made of small molecules.
Properties	High **melting** and **boiling points** because the strong covalent bonds between the atoms must be broken to melt or boil the substances. This requires a lot of energy. Solid at room temperature.	Low melting and boiling points because only the intermolecular forces need to be overcome to melt or boil the substances. This does not require a lot of energy as the intermolecular forces are weak. Normally gaseous or liquid at room temperature.

Most covalent structures do not conduct electricity because they do not have **delocalised electrons** or ions that are free to move to carry charge.

Ions

As well as sharing electrons, atoms can gain or lose electrons to give them a full outer shell. The number of protons is then different from the number of electrons. The resulting particle has a charge and is called an **ion**.

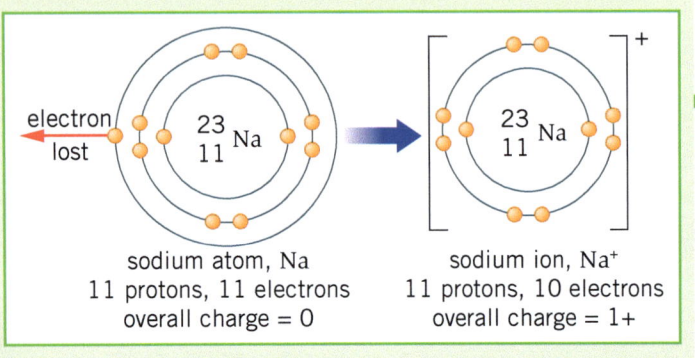

sodium atom, Na
11 protons, 11 electrons
overall charge = 0

sodium ion, Na$^+$
11 protons, 10 electrons
overall charge = 1+

Ionic bonding

When metal atoms react with non-metal atoms they **transfer** electrons to the non-metal atom (instead of sharing them).

Metal atoms lose electrons to become positive ions. Non-metal atoms gain electrons to become negative ions.

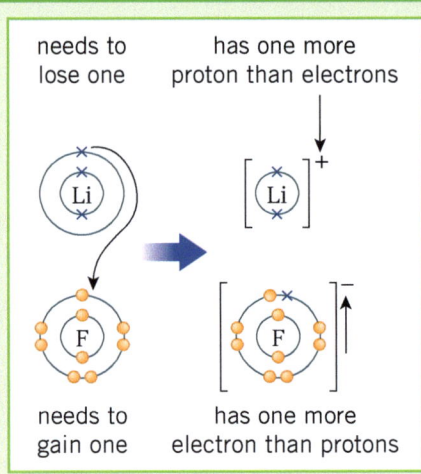

C2

Giant ionic lattice

Positive metal ions and negative non-metal ions are attracted to each other by the strong **electrostatic force of attraction**. This is called ionic bonding.

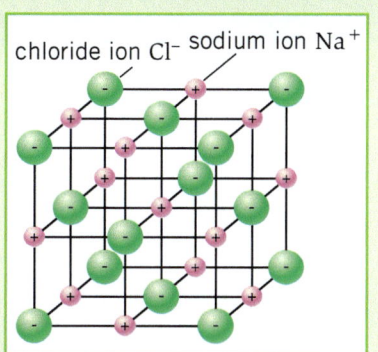

The electrostatic force of attraction works in all directions, so many billions of ions can be bonded together in a 3D structure called a **giant ionic lattice**.

The formula as an ionic substance can be worked out from the ratio of different ions in the bonding or lattice diagram, e.g. for every one chloride ion there's one sodium ion.

Melting points

Ionic substances have high melting points because the electrostatic force of attraction between oppositely charged ions is strong and so requires lots of energy to break.

Conductivity

Solid ionic substances do not conduct electricity because the ions are fixed in position and not free to carry charge.

When melted or dissolved in water, ionic substances do conduct electricity because the ions are free to move and carry charge.

Graphite

Graphite is a giant covalent structure.

Structure

Made only of carbon – each carbon atom bonds to three others, and forms hexagonal rings in layers.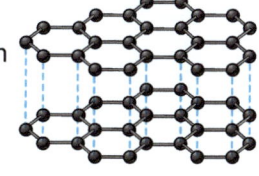
Each carbon atom has one spare electron, which is delocalised and therefore free to move around the structure.

Hardness

The layers can slide over each other because they are not covalently bonded. So graphite is softer than diamond, even though both are made only of carbon, as each atom in diamond has four covalent bonds.

Conductivity

The delocalised electrons are free to move through graphite, so can carry charges and allow an electrical current to flow. So graphite conducts electricity.

Fullerenes

- hollow cages of carbon atoms bonded together
- can be arranged as a sphere or a tube (called a **nanotube**)
- molecules held together by weak intermolecular forces, so can slide over each other
- conduct electricity

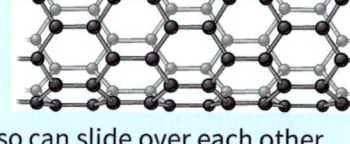

Nanoparticles

Nanoparticles are 1–100 nm in size.

Nanoparticles often have different properties to bulk materials of the same substance, caused by their **high surface area-to-volume ratio**. They are used in healthcare, electronics, cosmetics, and as catalysts. However, nanoparticles may be hazardous to health and to ecosystems, so it is important that they are researched further.

Graphene

Graphene consists of only a single layer of graphite. It's a strong material that can also conduct electricity. It can be used in composites and high-tech electronics.

Key Terms

Make sure you can write a definition for these key terms.

boiling point   conductivity   conductor   covalent bond   delocalised electrons   double bond   electrostatic force of attraction   fullerene   giant covalent   graphene   graphite   intermolecular forces   ionic   ion lattice   melting point

C2 Knowledge   15

Retrieval

Learn the answers to the questions below then cover the answers column with a piece of paper and write down as many as you can. Check and repeat.

C2 questions | Answers

#	Question	Answer
1	How are covalent bonds formed?	by atoms sharing electrons
2	How many electrons go into a covalent bond?	2 for a single bond, 4 for a double bond
3	Between which kinds of atoms does covalent bonding occur?	non-metals
4	What are the two main types of covalent structure?	giant covalent, small molecules
5	Describe the structure and bonding of a giant covalent substance.	billions of atoms bonded together by strong covalent bonds
6	Describe the structure and bonding of small molecules.	small numbers of atoms group together into molecules with strong covalent bonds between the atoms and weak intermolecular forces between the molecules
7	In general, what are the melting and boiling points of the two main types of covalent structure?	giant: high; small molecules: low;
8	Explain why giant covalent substances have high melting points.	strong covalent bonds between atoms require a lot of energy to break
9	Explain why small molecules have low melting points.	weak intermolecular force requires little energy to break
10	What is the name given to an atom that has gained or lost electrons?	an ion
11	Explain why large molecules have higher melting and boiling points than small molecules.	the intermolecular force is stronger
12	Why do most covalent substances not conduct electricity?	they do not have delocalised electrons or ions
13	Describe the structure and bonding in graphite.	each carbon atom is bonded to three others in hexagonal rings arranged in layers – it has delocalised electrons and weak intermolecular forces between the layers
14	Why can graphite conduct electricity?	the delocalised electrons can move through the graphite
15	Explain why graphite is soft.	layers are not bonded so can slide over each other
16	What is graphene?	one layer of graphite
17	Why do ionic compounds have high melting points?	there is strong electrostatic attraction between oppositely charged ions
18	When do ionic compounds conduct electricity?	when molten or in solution
19	If an atom loses one electron, what charge will its ion have?	+1
20	What do we call a large repeating structure of oppositely charged ions?	a giant ionic lattice

C2 Structure and bonding

Now use the questions below to check your knowledge from previous chapters.

C2

Previous questions | Answers

#	Question	Answer
1	What is an atom?	the smallest part of an element that can exist
2	What is an element?	a substance made from one type of atom
3	What is charge?	a fundamental property of particles
4	What are the three possible charges?	positive, negative, neutral
5	Where can you look up all the elements?	on the Periodic Table
6	What is the relative mass of a proton?	1
7	What is the relative mass of a neutron?	1
8	What is the relative mass of a electron?	0 (very small)
9	What do we call the change of state when we heat a liquid?	boiling
10	What is a compound?	substance made of more than one type of atom chemically joined together
11	Describe the structure of the atom.	a dense nucleus of protons and neutrons with orbiting electrons

(Put paper here)

 Maths Skills

Practise your maths skills using the worked example and practice questions below.

Plotting straight lines

When numerical data is plotted onto a graph you usually need to draw a line of best fit.

Sometimes this will be a straight line, but other times it will be a curve. You should draw whichever type of line fits the data.

Worked example

Early chemists carried out many experiments to work out the properties of different elements. One experiment was to heat a sample in oxygen and see how its mass changes depending on the mass of oxygen used. In one experiment, a scientist obtained the data below.

Mass of oxygen in g	Mass increase of element in g
5.0	2.1
10.0	4.0
15.0	6.2
20.0	8.1
25.0	9.8

This produces a graph with a **positive correlation** – as the value on the x-axis increases, so does the value on the y-axis.

With a **negative correlation** the value on the x-axis increases, the value on the y-axis decreases.

positive correlation +1

Practice

In another experiment, scientists obtained the data below.

Mass of oxygen in g	Mass increase of element in g
0.0	0.0
4.0	5.2
8.0	10.1
12.0	14.7
16.0	19.8
20.0	25.1

1. Using graph paper, draw a graph for these data and include a straight line of best fit.
2. Does your graph show a positive or negative correlation?
3. In another experiment, scientists looked at how the mass of a 5.0 g element increased as it was heated. Where does the line of best fit start on this graph, compared to on your graph?

C2 Retrieval 17

Practice

Exam-style questions

01 Silicon dioxide has a giant covalent structure. It has a high melting point and does not conduct electricity.

01.1 Draw **one** line from each property to the explanation of the property. **[2 marks]**

Property	Explanation
	strong intermolecular forces of attraction
high melting point	there are no charged particles free to move
does not conduct electricity	strong covalent bonds
	electrons are free to move

> **Exam Tip**
> Don't be tempted to draw four lines just because there are four boxes on the right.
> Only draw **two** lines in total, one from each of the boxes on the left.

01.2 **Figure 1** shows three suggested structures of silicon dioxide, SiO_2.

Figure 1

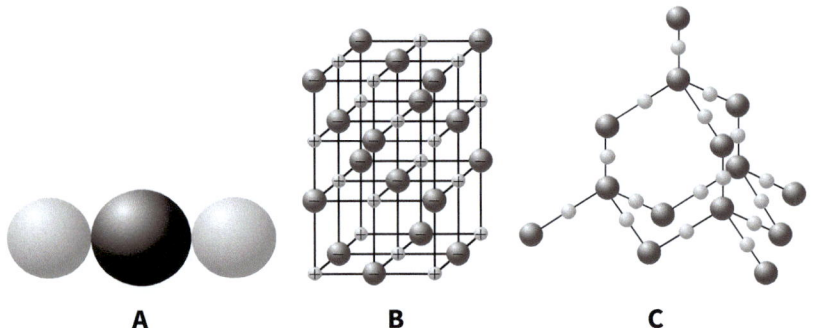

A   B   C

Identify which structure is the correct structure of silicon dioxide. **[1 mark]**

_____

01.3 Silicon dioxide contains the elements silicon and oxygen.
Table 1 shows some properties of silicon and oxygen.

Table 1

	Melting point in °C	Boiling point in °C	Conducts electricity
oxygen	−218.8	−183	no
silicon	1414.0	3265	yes

Use **Table 1** and the Periodic Table to identify the type of structure of oxygen and silicon. **[2 marks]**

oxygen gas _____

silicon _____

> **Exam Tip**
> Look at the position of oxygen on the Periodic Table to help you identify the bonding in oxygen gas.

18   C2 Structure and bonding

02 Phosphorus is a Group 5 element. It reacts with hydrogen to produce a compound called phosphine.

02.1 The electronic structure of phosphorus is 2,8,5.

Complete the dot and cross diagram in **Figure 2** to show the covalent bonding in a molecule of phosphine, PH_3.

You should show only the electrons in the outer shells. **[2 marks]**

Figure 2

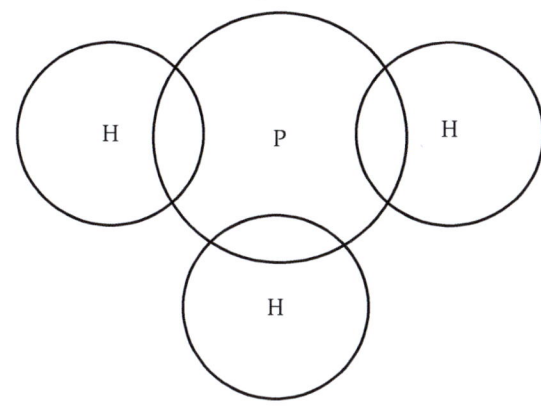

> **Exam Tip**
>
> Use crosses for hydrogen and dots for phosphorus. The groups they are in on the Periodic Table will help you work out how many electrons are in each outer shell.

02.2 Name the type of bond or force overcome when liquid phosphine boils. **[1 mark]**

_____

02.3 **Table 2** shows the boiling points of the compounds formed between hydrogen and the elements of Group 5 of the Periodic Table.

Table 2

Formula of compound	Radius of central atom in nm	Boiling point in °C
PH_3	0.110	−88
AsH_3	0.121	−55
SbH_3	0.141	−17
BiH_3	0.152	16

Identify the state that phosphine is in at room temperature. **[1 mark]**

_____

02.4 Describe the trend shown in **Table 2**.

Suggest a reason for this trend. **[2 marks]**

_____

_____

_____

> **Exam Tip**
>
> Increase or decrease isn't enough – an increase or decrease in *what*?

03 Graphene is a single layer of graphite. It can be represented using a ball and stick model.

03.1 The ball and stick model is not a true representation of the structure of graphene. Give **one** reason why. **[1 mark]**

03.2 Explain why graphene conducts electricity. **[1 mark]**

03.3 Graphene is made up of carbon atoms only. One carbon atom has a mass of 1.99×10^{-23} g. A scientist has a sheet of graphene of mass 0.240 g. Calculate the number of carbon atoms in the sheet of graphene. Give your answer to three significant figures. **[3 marks]**

> **Exam Tip**
> Don't be worried by very small or very big numbers – first check you can put them into your calculator correctly, then carry out the calculations.

04 Compare the physical properties of diamond and graphite. In your answer, use ideas about bonding to explain the differences in properties. **[6 marks]**

> **Exam Tip**
> Ensure you link each property to the feature of its bonding that gives rise to that property.

05 **Table 3** shows some properties of three elements, **X**, **Y**, and **Z**. The letters are *not* the chemical symbols of the elements.

Table 3

Element	Melting point in °C	Does the element conduct electricity?
X	−219	no
Y	−101	no
Z	very high	yes

05.1 Identify which element in **Table 3** could be carbon, in the form of graphite. **[1 mark]**

05.2 One of the elements in **Table 3** is chlorine. Draw a dot and cross diagram to show the covalent bonding in a molecule of chlorine, Cl_2. You should show only the electrons in the outer shells. **[2 marks]**

05.3 One of the elements in **Table 3** is oxygen. Draw a dot and cross diagram to show the covalent bonding in a molecule of oxygen, O_2. You should show only the electrons in the outer shells. **[2 marks]**

> **Exam Tip**
> The bonding in oxygen gas is a little bit more complicated than in chlorine gas.

05.4 An oxygen atom is smaller than a chlorine atom. Deduce the letter of the element in **Table 3** that represents chlorine. Justify your choice. **[2 marks]**

06 Figure 3 shows the ball and stick model of a compound, **X**.

Figure 3

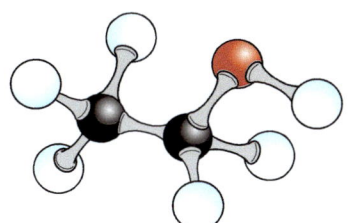

06.1 Predict **two** physical properties of compound **X**. Explain why compound **X** has each of the properties you predicted. **[4 marks]**

06.2 In **Figure 3**, the different coloured balls represent atoms of different elements:

black = carbon   white = hydrogen   red = oxygen.

Deduce the molecular formula of compound **X**. **[1 mark]**

06.3 A sample of compound **X** contains 6.02×10^{23} molecules. Calculate the number of hydrogen atoms in the sample. **[1 mark]**

> **Exam Tip**
>
> Tick off the balls in the diagram once you have counted them, to make sure you don't count some twice.

06.4 The boiling point of compound **X** is 78 °C and the melting point is −114 °C. Identify the state of compound **X** at 25 °C. **[1 mark]**

07 Hydrocarbons are compounds that are made up of carbon atoms and hydrogen atoms only. **Table 4** gives some data on two hydrocarbons.

Table 4

Name of compound	Ball and stick model of molecule	Melting point in °C	Boiling point in °C
methane		−182	−162
hexane		−96	69

> **Exam Tip**
>
> You might not be used to seeing compounds drawn like this, but in reality large organic molecules are rarely sitting around in neat straight lines.

In the ball and stick models, dark grey spheres represent carbon atoms and white spheres represent hydrogen atoms.

07.1 Write the molecular formula of hexane. **[1 mark]**

07.2 Draw a dot and cross diagram to show the covalent bonding in a molecule of methane, CH_4. You should show only the electrons in the outer shells. **[2 marks]**

07.3 Draw the displayed formula of methane. In the formula, represent each atom with its chemical symbol and each single covalent bond with a line. **[1 mark]**

> **Exam Tip**
>
> If you're not sure, start by drawing a stick diagram, then a diagram with five overlapping circles, and then add the electrons.

07.4 Use your own knowledge *and* the data in **Table 4** to compare the physical properties of methane and hexane at room temperature, 20 °C. **[6 marks]**

07.5 Explain, in terms of the forces between molecules, why hexane has a higher boiling point that methane. **[2 marks]**

08 An ion has the formula $^{69}_{31}\text{Ga}^{3+}$.

08.1 Give the number of protons in the ion. **[1 mark]**

08.2 Give the number of neutrons in the ion. **[1 mark]**

08.3 Give the number of electrons in the ion. **[1 mark]**

! **Exam Tip**

The charge on this ion will affect the number of electrons, not the number of protons.

08.4 Give the name of **one** other element that is in the same group of the Periodic Table as gallium. **[1 mark]**

09 **Table 5** shows some properties of two oxides.

Table 5

Compound	Formula	Boiling or sublimation temperature in °C
carbon dioxide	CO_2	sublimes at −79
silicon dioxide	SiO_2	boils at 2230

09.1 Draw a dot and cross diagram for carbon dioxide. You should show only the electrons in the outer shells. **[2 marks]**

09.2 Explain the difference in the boiling and sublimation temperatures shown in **Table 5**. **[3 marks]**

! **Exam Tip**

Sublimation is when a compounds turns from a solid to a gas, without becoming a liquid.

10 **Table 6** shows the melting points of some ionic compounds.

Table 6

Compound	Melting point in °C
calcium oxide	2572
calcium sulfide	2525
calcium bromide	730
magnesium oxide	2852
magnesium sulfide	2000
magnesium bromide	711
sodium oxide	1132
sodium sulfate	884
sodium bromide	747

10.1 Describe the pattern between the charges of the ions in a compound and the melting point of the compound. **[6 marks]**

! **Exam Tip**

It might help to note the charges of each ion next to the table so that it is easy to compare the charges to the melting points.

10.2 Explain the general pattern observed in **Table 6**. **[2 marks]**

11 **Figure 4** shows the outer electrons in an atom of magnesium and in an atom of bromine. Magnesium is in Group 2 of the Periodic Table and bromine is in Group 7.

Figure 4

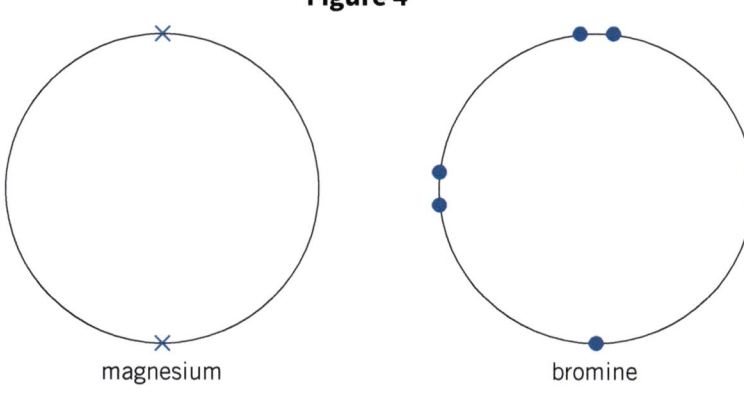

magnesium   bromine

11.1 Magnesium and bromine form an ionic compound. Describe what happens to the electrons when magnesium reacts with bromine. **[3 marks]**

11.2 Give the formulae of the ions formed in the reaction between magnesium and bromine. **[2 marks]**

11.3 Give the chemical formula of the compound formed. **[1 mark]**

11.4 Predict the physical properties of the compound formed. **[3 marks]**

> **Exam Tip**
>
> Bromine only needs one more electron to get a full outer shell and magnesium has two electrons to give away. Think carefully about the ratio of bromine to magnesium.

12 This question is about two compounds: caesium oxide, Cs_2O, and dichlorine monoxide, Cl_2O.

12.1 Draw a dot and cross diagram for dichlorine monoxide. **[2 marks]**

12.2 Describe the difference in bonding between the two compounds. **[5 marks]**

12.3 Compare the physical properties of caesium oxide and dichlorine monoxide. Explain your predictions. **[4 marks]**

12.4 The melting point of caesium oxide is 490 °C. The melting point of barium oxide is 1923 °C. Barium is in Group 2 of the Periodic Table but the same period as caesium. Suggest why barium oxide has a significantly higher melting point than caesium oxide. **[2 marks]**

> **Exam Tip**
>
> Before you start, mark which compound is ionic and which is covalent, this will prevent you from getting confused later.

> **Exam Tip**
>
> Even if these compounds are unfamiliar to you, the same rules apply as to any other compound. Approach the questions logically and you'll be fine.

13 **Figure 5** shows the outer electrons of a potassium atom and an oxygen atom.

Figure 5

13.1 Draw a dot and cross diagram for the ionic compound formed when oxygen reacts with potassium. **[3 marks]**

13.2 Describe how the ions are bonded together in potassium oxide. **[3 marks]**

13.3 The melting point of potassium oxide is 740 °C. The melting point of oxygen is −218 °C. Explain why the melting point of potassium oxide is much higher than that of oxygen. **[5 marks]**

13.4 The melting point of potassium is 63.5 °C. Give **one** conclusion that can be made about metallic and ionic bonding using this data and the data from **13.3**. **[1 mark]**

> **Exam Tip**
> Remember ionic compounds have square brackets – not overlapping circles.

14 **Figure 6** shows the structure of part of an ionic compound.

Figure 6

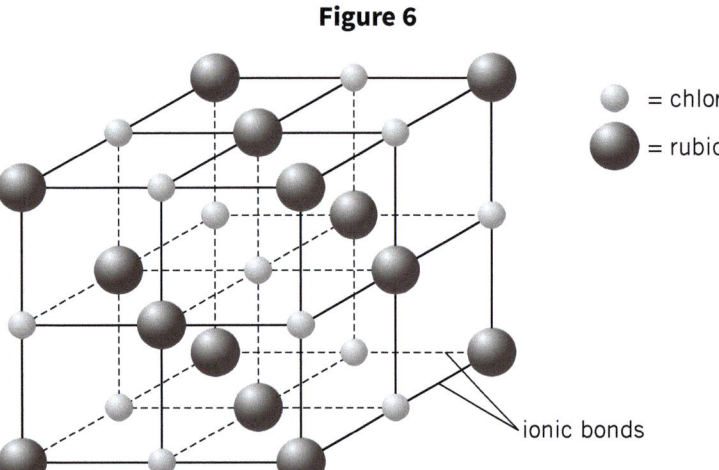

= chloride ion
= rubidium ion
ionic bonds

14.1 Suggest the most likely empirical formula of the ionic compound. **[1 mark]**

14.2 Suggest how **Figure 6** could be improved so that it shows the exact empirical formula of the compound. Explain your suggestion. **[2 marks]**

14.3 List **two** incorrect assumptions made about the rubidium chloride in the model shown in **Figure 6**. **[2 marks]**

> **Exam Tip**
> There are a large number of atoms shown in the diagram, but this question is asking you for the formula that shows the ratio of atoms in its simplest form. This is called the empirical formula.

15 Silver nanoparticles can be used to kill disease-causing bacteria. The nanoparticles enter a bacterial cell through its wall and membrane.

15.1 Suggest **two** reasons that might explain why silver nanoparticles are more effective than bulk silver at killing bacteria. **[2 marks]**

15.2 Suggest **one** possible risk of using silver nanoparticles in medicines for humans. **[1 mark]**

Exam Tip

The main body of the question is short but it provides lots of information to help answer these questions; don't ignore it.

15.3 **Figure 7** shows the percentage of bacteria that survived after coming in contact with different amounts of silver nanoparticles.

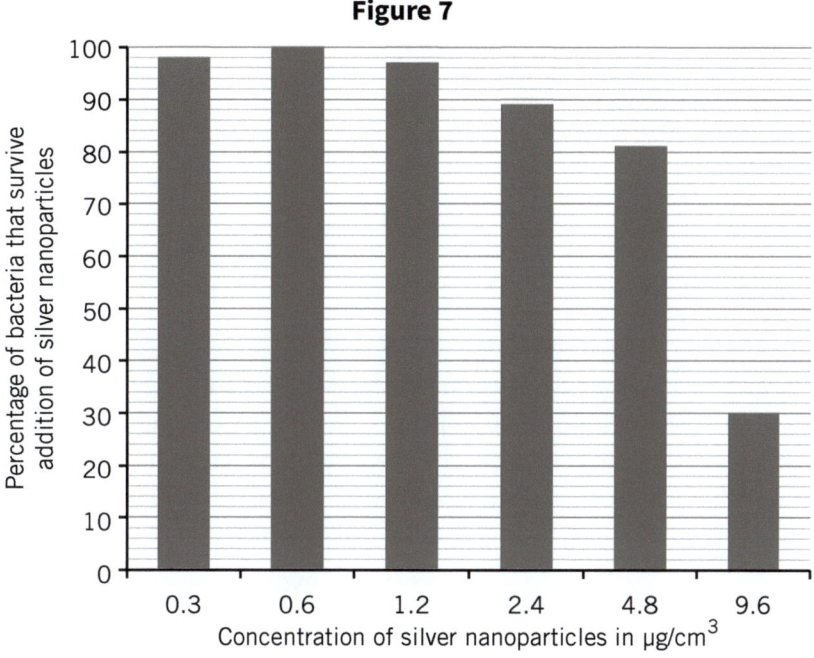

Figure 7

Describe the pattern shown on the bar chart. **[2 marks]**

15.4 Give **two** other uses of nanoparticles. **[2 marks]**

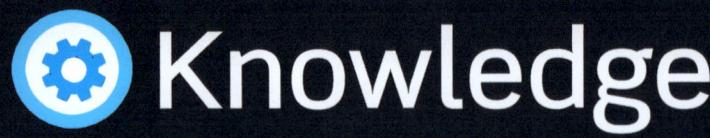 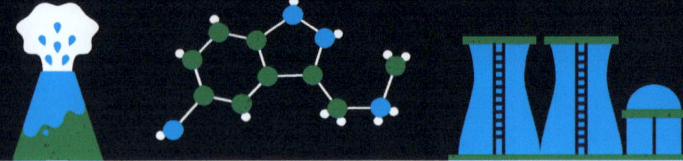

C3 The Periodic Table

The modern Periodic Table

There are over 100 elements listed on the **Periodic Table**. The design of the Periodic Table has changed over time and is based on the work of Mendeleev. The modern Periodic Table:

- is ordered by atomic number
- has no gaps
- groups the elements by the number of electrons in the outer shell.

Elements with the same number of electrons in their outer shell have similar chemical properties. So you can predict the properties of elements in a group.

Group 0

Elements in Group 0 are called the **noble gases**. Noble gases have the following properties:

- full outer shells with eight electrons, so do not need to lose or gain electrons (helium has 2 electrons in its outer shell)
- are very **unreactive** (**inert**) so exist as single atoms as they do not bond to form molecules
- boiling points that increase down the group

Group 1 elements

Group 1 elements are low-density metals that form white ionic compounds. They react with oxygen, chlorine, and water:

lithium + oxygen → lithium oxide
lithium + chlorine → lithium chloride
lithium + water → lithium hydroxide + hydrogen

Group 1 elements are called **alkali metals** because they react with water to form an alkali (a solution of their metal hydroxide).

Group 1 properties

Group 1 elements all have one electron in their outer shell. They are very reactive because they only need to lose one electron to react. Reactivity increases down Group 1 because as you move down the **group**:

- the atoms increase in size
- the outer electron is further away from the nucleus, and there are more shells shielding the outer electron from the nucleus
- the electrostatic attraction between the nucleus and the outer electron is weaker
- so it is easier to lose the one outer electron.

The melting points and boiling points decrease down Group 1.

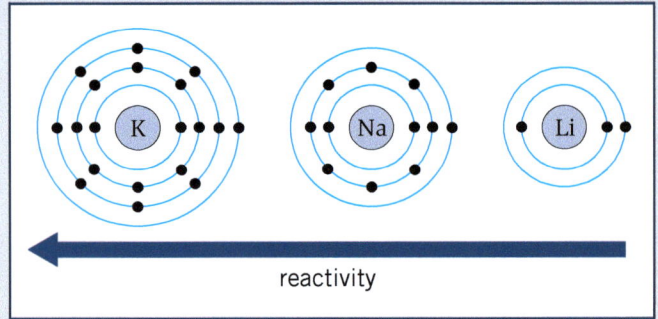

Key Terms

Make sure you can write a definition for these key terms.

alkali metals	chemical properties	displacement	groups	halogens	inert
noble gas	Periodic Table	reactivity	transition metals	unreactive	

C3

Group 7 elements

Group 7 elements are called the **halogens**. They are non-metals that exist as molecules made up of pairs of atoms. The melting points and boiling points of the halogens increase down the group. The reactivity decreases down the group. When halogens form ions they are called **halide ions** and have a charge of –1.

Name	Formula	State at room temperature
fluorine	F_2	gas
chlorine	Cl_2	gas
bromine	Br_2	liquid
iodine	I_2	solid

Group 7 displacement

More reactive Group 7 elements can take the place of less reactive ones in a compound. This is called **displacement**. For example, fluorine displaces chlorine as it is more reactive:

fluorine + potassium chloride → potassium fluoride + chlorine

Group 7 reactivity

Reactivity decreases down Group 7 because as you move down the group:
- the atoms increase in size
- the outer shell is further away from the nucleus, and there are more shells between the nucleus and the outer shell
- the electrostatic attraction from the nucleus to the outer shell is weaker
- so it is harder to gain the one electron to fill the outer shell.

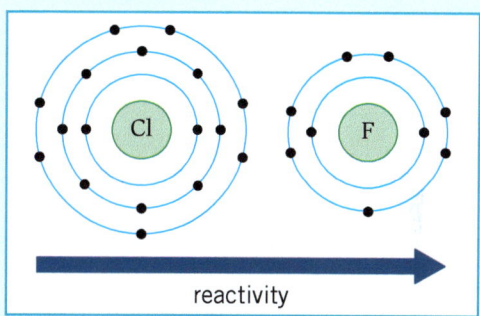

reactivity

Transition metals

The transition metals can be found in the middle of the Periodic Table. Their properties are similar to each other, but are different from the Group 1 metals.

Property	Group 1	Transition metals
melting point	relatively low (e.g., sodium melts at 98 °C)	relatively high (e.g., iron melts at 1538 °C)
density	relatively low	relatively high (e.g., iron is almost ten times as dense as sodium)
strength	relatively low	relatively high
hardness	relatively low (e.g., quite easy to scratch)	relatively high (e.g., quite hard to scratch)
reactivity with oxygen, water, and halogens	react easily with oxygen, water, and halogens	react very slowly, if at all, with oxygen, water, and halogens

Catalysts

Transition metals are very useful as catalysts – they increase the rate of a reaction without being used up.

Coloured compounds

When transition metals form compounds, they often take on a colour. For example, chromium(III) oxide is green and cobalt(II) sulfate is red.

Ions

Group 1 metals only form 1+ ions and Group 2 metals only form 2+ ions. However, most transition metals can form many differently charged ions, for example, copper can form Cu^+ or Cu^{2+} ions, and manganese can form Mn^{2+}, Mn^{3+}, Mn^{4+}, Mn^{6+}, or Mn^{7+} ions. In a compound, the charge is given by roman numerals in brackets. For example, manganese(II) sulfate contains the Mn^{2+} ion.

C3 Knowledge

Retrieval

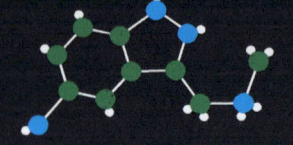 

Learn the answers to the questions below then cover the answers column with a piece of paper and write as many as you can. Check and repeat.

C3 questions | Answers

#	Question	Answer
1	How many electrons are in the outer shell of the group 7 elements?	7
2	What group number is given to the inert elements?	0
3	Why do elements in a group have similar chemical properties?	they have the same number of electrons in their outer shell
4	What name is given to the Group 1 elements?	alkali metals
5	Why are the alkali metals called alkali metals?	they are metals that react with water to form an alkali
6	Give the general equations for the reactions of alkali metals with oxygen, chlorine, and water.	metal + oxygen → metal oxide metal + chlorine → metal chloride metal + water → metal hydroxide + hydrogen
7	How does the reactivity of the alkali metals change down the group?	increases
8	Which is the least reactive alkali metal?	lithium
9	Why does the reactivity of the alkali metals increase down the group?	they are larger atoms, so the outermost electron is further from the nucleus, meaning there are weaker electrostatic forces of attraction and more shielding between the nucleus and outer electron, and it is easier to lose the electron
10	What name is given to the Group 7 elements?	halogens
11	Give the formulae of the first four halogens.	F_2, Cl_2, Br_2, I_2
12	How do the melting points of the halogens change down the group?	increase
13	How does the reactivity of the halogens change down the group?	decrease (less reactive)
14	What is a displacement reaction?	when a more reactive element takes the place of a less reactive one in a compound
15	What name is given to the Group 0 elements?	noble gases
16	Why are the noble gases inert?	they have full outer shells so do not need to lose or gain electrons
17	How do the melting points of the noble gases change down the group?	increase

C3 The Periodic Table

Now use the questions below to check your knowledge from previous chapters.

C3

Previous questions | Answers

	Previous questions	Answers
1	What do we call the state change when a gas cools?	condensation
2	In general, what are the melting and boiling points of the two main types of covalent structure?	giant: high, small molecules: low
3	Explain why giant covalent substances have high melting points.	strong covalent bonds between atoms require a lot of energy to break
4	Explain why small molecules have low melting points.	weak intermolecular force requires little energy to break
5	What is the name given to an atom which has gained or lost electrons?	an ion
6	An atom loses an electron, what charge will its ion have?	+1
7	What do we call a large repeating structure of oppositely charged ions?	a giant ionic lattice

(Put paper here)

Maths Skills

Practise your maths skills using the worked example and practice questions below.

Plotting curves

Remember that you need to draw a line of best fit when numerical data are plotted on a graph.

Some data will need a curved line of best fit, rather than a straight one.

It is important to remember that you draw the line that best fits the data.

Revision Tip

Ignore anomalous results when you draw a line or curve of best fit.

Worked example

The alkanes have the boiling points given below. Plot the data on a graph, and draw an appropriate line of best fit.

Number of carbon atoms	Boiling point in °C
1	−162.0
2	−89.0
3	−42.0
4	0.0
5	36.0
6	69.0

The graph has a curved line of best fit and shows a positive correlation – as the number of carbon atoms increases, so does the boiling point.

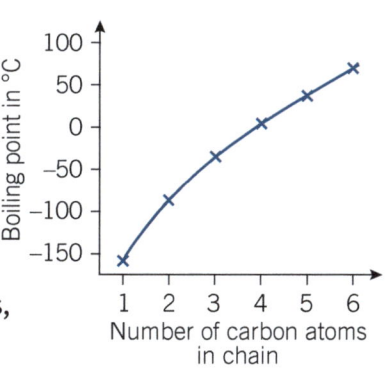

Practice

A different group of hydrocarbons have the boiling points given below.

Number of carbon atoms	Boiling point in °C
4	5.1
5	44.7
6	72.8
8	106.0
9	112.4
10	115.6

1 Plot a graph of these results. Draw an appropriate line of best fit.

2 Use your graph from **1** to predict the boiling point of a hydrocarbon in this group with seven carbon atoms.

3 Does your graph show a positive or negative correlation?

C3 Retrieval 29

Practice

Exam-style questions

01 Rubidium is in Group 1 of the Periodic Table.

01.1 Is rubidium a compound, metal, or a non-metal? **[1 mark]**

_____

01.2 Identify the products when rubidium reacts with water. **[1 mark]**
Tick **one** box.

rubidium oxide and oxygen ☐

rubidium hydroxide and oxygen ☐

rubidium chloride and hydrogen ☐

rubidium hydroxide and hydrogen ☐

> **Exam Tip**
> The formula of water might give you a clue.

01.3 Rubidium also reacts with oxygen.
Write the word equation for this reaction. **[1 mark]**

_____

01.4 Sodium is another element in Group 1 of the Periodic Table. Sodium reacts with bromine.
Complete the balanced symbol equation for the reaction between sodium and bromine. **[2 marks]**

_____ Na(_____) + Br_2(_____) → _____ NaBr(s)

> **Exam Tip**
> All you need to add is the numbers in front of the elements and compounds and the state symbols. Don't add in any other compounds.

01.5 Rubidium also reacts with bromine.
Explain the difference in the reactivity of sodium and rubidium with chlorine. **[4 marks]**

_____
_____
_____
_____

02 The columns of the Periodic Table are called groups. The elements in a group have similar properties.

02.1 Draw **one** line from each group to a property of the elements in this group. **[3 marks]**

> **Exam Tip**
> Only draw three lines here, even though there are five boxes on the right-hand side.

02.2 Explain the difference in the trend in reactivity down Group 1 and Group 7. **[6 marks]**

_____
_____
_____
_____
_____
_____

> **Exam Tip**
> The answer to **02.2** needs to have two sections.
> Make it clear which group you are talking about in each section.

02.3 Explain the reactivity of Group 0. **[2 marks]**

_____
_____

03 Xenon is in Group 0 and Period 5 of the Periodic Table. Under extreme conditions, xenon will react with fluorine.

03.1 Explain why xenon and fluorine are able to react. **[3 marks]**

03.2 When xenon reacts with fluorine, the xenon atom is able to have 12 electrons in its outer shell. Complete the dot and cross diagram in **Figure 1** to show the product of the reaction between xenon and fluorine. **[2 marks]**

03.3 Identify the type of bonding in xenon tetrafluoride. **[1 mark]**

Figure 1

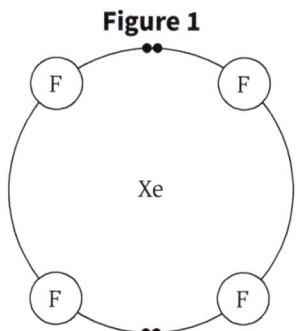

04 A teacher demonstrated the reaction of sodium with water. This is the method used:

1. Fill a big glass trough with water.
2. Use tongs to take a lump of sodium out of the oil in its storage bottle.
3. Cut off a small piece of sodium.
4. Put the bigger lump of sodium back in its storage bottle.
5. Use tongs to place the small piece of sodium on the surface of the water.

04.1 Identify which group of the Periodic Table sodium is in. Choose **one** answer.
Group 1   Group 2   Group 3   Group 4   **[1 mark]**

04.2 Suggest a reason for step **4**. **[1 mark]**

04.3 Suggest a reason for using tongs in step **5**. **[1 mark]**

04.4 In step **5**, the sodium does not start reacting with the water immediately. Suggest an improvement to step **3** to make the reaction start more quickly. **[1 mark]**

04.5 Name the gas made in the reaction. **[1 mark]**

04.6 Describe how the teacher could show that one of the products of the reaction makes an alkaline solution in water. **[2 marks]**

04.7 Explain why the reaction of lithium with water is less vigorous than the reaction of sodium with water. In your answer, include the electronic structures of lithium and sodium. **[3 marks]**

04.8 Caesium is an element near the bottom of Group 1. Predict the observations on adding caesium to water. **[1 mark]**

> **Exam Tip**
> Being able to read a method and suggest improvements is an important skill in science.

> **Exam Tip**
> First decide if caesium is more or less reactive than sodium.

05 A student carried out some reactions of halogens with solutions of potassium chloride, potassium bromide, and potassium iodide. The solutions were labelled **X**, **Y**, and **Z**. **Table 1** shows the student's results.

Table 1

Reacted with	Solution X	Solution Y	Solution Z
chlorine water	yellow solution formed	no change observed	brown solution formed
bromine water	no change observed	no change observed	brown solution formed
iodine water	no change observed	no change observed	no change observed

Deduce the identities of solutions **X**, **Y**, and **Z**. Justify your decisions. Use electronic structures to suggest an explanation for one of the reactions that occurs. **[6 marks]**

06 Four pairs of substances are reacted together:

    A   lithium and bromine   **C**   sodium and bromine

    B   lithium and fluorine   **D**   sodium and fluorine

Predict which pair of substances has the most vigorous reaction. Explain your prediction. **[6 marks]**

Exam Tip

For this answer you'll have to refer to the locations of the elements on the Periodic Table and their structures.

07 **Figure 2** shows the electronic structures of some Group 0 elements. Each is labelled with a letter. The letters are not the chemical symbols of the elements.

Figure 2

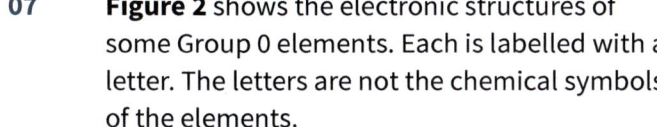

07.1 Give the name used for the elements in Group 0 of the Periodic Table. **[1 mark]**

07.2 Give the letter that represents a helium atom in the Periodic Table. **[1 mark]**

07.3 Give the letter of the atom of the element in **Figure 2** that has the lowest boiling point. **[1 mark]**

07.4 Draw the electronic structure of a neon atom. **[1 mark]**

Exam Tip

You get no marks for perfect circles, so try not to spend a long time drawing.

07.5 Explain why Group 0 elements do not readily form molecules. **[2 marks]**

08 **Figure 3** shows the electronic structures of four atoms. Each atom is labelled with a letter. The letters are not the chemical symbols of the elements.

Figure 3

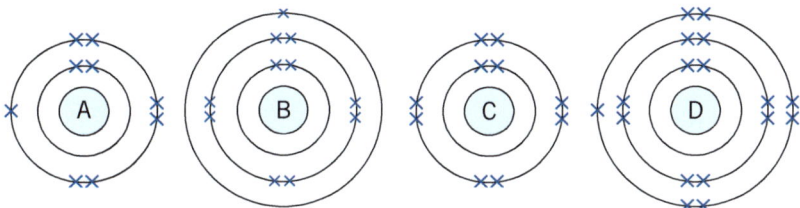

Exam Tip

Looking at the electrons in the outer shells of these elements is the key to answering the questions.

08.1 Give the letter of the atom of a Group 1 element. **[1 mark]**

08.2 Give the letter of the atom of an unreactive element and explain why this element is unreactive. **[2 marks]**

08.3 Give the letters of two atoms of elements that are in the same group of the Periodic Table. **[1 mark]**

09 Figure 4 shows the electronic structures of the atoms of three Group 2 elements.

Figure 4

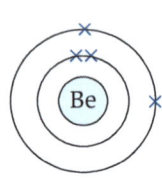
beryllium

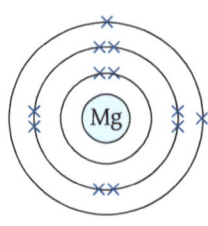
magnesium

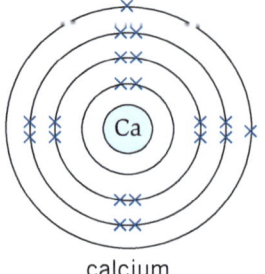
calcium

09.1 Predict how the reactivity of the Group 2 elements changes from the top to the bottom of the group. Justify your prediction by comparing the Group 2 electronic structures to the electronic structures of Group 1. **[4 marks]**

> **Exam Tip**
>
> In written text you can use a shorter notation to refer to the electronic structures. For example, Be would be 2,2, Mg would be 2,8,2, and Ca would be 2,8,8,2.

09.2 Describe the structure of calcium. **[3 marks]**

09.3 Magnesium reacts with steam, but does not react with cold water. Is magnesium more or less reactive than sodium? Give a reason for your answer. **[2 marks]**

09.4 Magnesium reacts with chlorine in a similar way to sodium. Give the chemical formula of the product formed when magnesium reacts with chlorine. **[1 mark]**

> **Exam Tip**
>
> You'll need to know the formula for ions of magnesium and chlorine to work out the answer.
>
> You can either learn these or work them out from the Periodic Table.

10 The word equations for three reactions are given below.

Reaction 1 hydrogen + fluorine → hydrogen fluoride
Reaction 2 hydrogen + bromine → hydrogen bromide
Reaction 3 iron + bromine → iron bromide

10.1 Draw the electronic structure of the product of reaction **1**. **[2 marks]**

10.2 Explain why reaction **1** is more vigorous than reaction **2**. In your explanation, include the electronic structures of the halogens involved in the reactions. **[6 marks]**

> **Exam Tip**
>
> Before you answer, look up each of the elements in this question on the Periodic Table.

10.3 Predict whether the product of reaction **2** or reaction **3** melts at the higher temperature. Justify your prediction. **[3 marks]**

11.1 Give the name of Group 1 in the Periodic Table. **[1 mark]**

11.2 Elements in the same group in the periodic table have similar chemical properties. Select which answer best describes this. **[1 mark]**

  A They have the same number of electrons in the shell nearest the nucleus.

  B They have the same number of electron shells.

  C They have the same number of electrons in the shell furthest from the nucleus.

  D They have the same number of electrons.

C3 The Periodic Table

11.3 Identify whether Group 1 elements are metals or non-metals. **[1 mark]**

11.4 Lithium is a Group 1 element. Lithium reacts with chlorine, a Group 7 element. The product is lithium chloride. Caesium is another Group 1 element, and bromine is another Group 7 element. Name the product when caesium reacts with bromine. **[1 mark]**

> **Exam Tip**
> The location of metals and non-metals on the Periodic Table will not be given to you in the exam, you need to know this.

12 A student has a dilute potassium chloride solution.

12.1 Identify which image in **Figure 5** shows the correct particle diagram for dilute potassium chloride solution. Choose **one** answer. **[1 mark]**

Figure 5

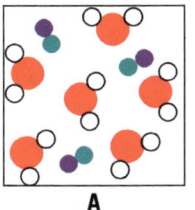 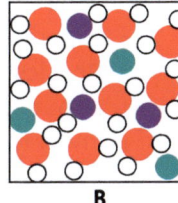 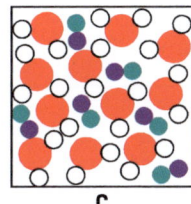 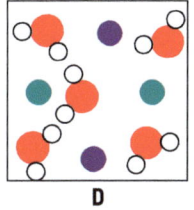

A   B   C   D

> **Exam Tip**
> The key phrase in this question is 'potassium chloride solution'. This means there will also be water in the substance.

12.2 Write the electronic structure of the potassium ion in potassium chloride. **[1 mark]**

12.3 Describe the structure and bonding in potassium chloride. **[4 marks]**

12.4 Give the reason why potassium chloride solution can conduct electricity. **[1 mark]**

12.5 Describe a method by which potassium chloride can be separated from the solution. Your method should result in both potassium chloride and water being collected. **[6 marks]**

13 Data for two metals are given in **Table 2**. Each metal is represented by a letter. The letters are **not** the chemical symbols of the metals.

Table 2

Metal	Reaction with oxygen	Colour of its compound with chlorine	Does it conduct electricity?	Observations on placing in water
A	when exposed to air, immediately forms a white coating	white	yes	fizzes, forming bubbles and an alkaline solution
B	small pieces burn in air; reacts very slowly with oxygen in the air at room temperature	brown	yes	over several days, forms a brown flaky substance

One metal in **Table 2** is a transition metal and one is in Group 1 of the Periodic Table.

13.1 Identify which physical property in **Table 2** *cannot* be used to distinguish between a Group 1 metal and a transition metal. **[1 mark]**

13.2 Identify which metal from **Table 2** is iron. Justify your decision. **[3 marks]**

13.3 Two oxides of iron are FeO and Fe_2O_3. Explain why the fact that iron forms two oxides is evidence that iron is a transition metal. **[2 marks]**

14 **Figure 6** shows apparatus to measure hardness. A scientist uses the apparatus in **Figure 6** to compare the hardness of three different metals. This is the method used:

1. Push the ball down with a force of 30 000 N.
2. Hold for 5 seconds.
3. Remove the ball.
4. Measure the diameter of the indentation on the surface of the metal.
5. Repeat with two different metals.

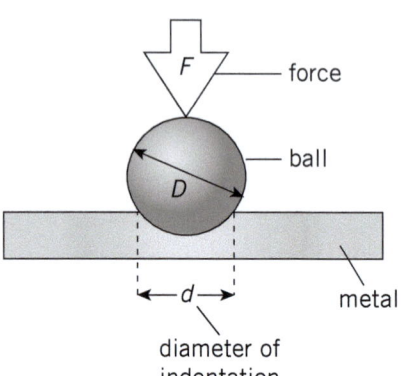

Figure 6

> **Exam Tip**
> You might think this experiment belongs in a Physics question; this is just about applying what you know to a new context so don't let that put you off!

14.1 Identify the independent and dependent variables in the experiment. **[2 marks]**

14.2 Explain why the force used is the same for all three metals. **[1 mark]**

14.3 Suggest why the ball in **Figure 6** is made from a very hard material. **[1 mark]**

14.4 Suggest **one** improvement that would reduce the effect of random errors in measuring the diameter of the indentation. **[1 mark]**

14.5 The scientist's results are shown in **Table 3**.

> **Exam Tip**
> Go over the method and pick out the independent and dependent variables.

Table 3

Substance	Diameter of indent in mm
A	2
B	6
C	7

Identify which metal is mostly likely to be the transition metal. Explain your answer. **[3 marks]**

> **Exam Tip**
> The first thing to do is to think about the properties of transition metals.

15 This question is about the elements in Group 0 of the Periodic Table.

15.1 Give the general name for the elements in Group 0. **[1 mark]**

15.2 An argon atom has 18 electrons. Write the electronic structure for argon. **[1 mark]**

15.3 Explain why the Group 0 elements are unreactive. **[1 mark]**

36 C3 The Periodic Table

15.4 Describe the trend in boiling points from top to bottom of Group 0.
[1 mark]

16 A teacher demonstrates the reaction of sodium with chlorine. This is the method used:
  1 Heat a small piece of sodium.
  2 Fill a gas jar with chlorine.
  3 Place the hot sodium in the gas jar of chlorine.

16.1 Suggest **two** safety precautions the teacher should take. **[2 marks]**

16.2 Explain an improvement to the order in which the steps above are carried out. **[2 marks]**

16.3 Draw a dot and cross diagram of the product of the reaction.
[2 marks]

16.4 Describe the bonding in the product. **[3 marks]**

16.5 Explain the difference in the conductivity of electricity between sodium, chlorine, and the product of the reaction. **[6 marks]**

16.6 Predict **one** difference in the observations made if chlorine was replaced by bromine. **[1 mark]**

> **! Exam Tip**
> Think about the substances being used in the reaction; they might give you a clue about the safety precautions required.

17 This question is about Group 1 elements.

17.1 Complete the sentence. **[1 mark]**
The elements in Group 1 are all **metals / non-metals**.

17.2 The reactivity of the Group 1 elements changes as you move down the group. Put these elements in order from least reactive to most reactive. **[3 marks]**

| caesium | sodium | rubidium | lithium | potassium |

17.3 The elements in Group 1 undergo a number of reactions. Draw **one** line from each pair of reactants to the correct products. **[3 marks]**

Reactants   **Products**

sodium + chlorine   sodium oxide

                         sodium chloride + hydrogen

sodium + water   sodium chloride

                         sodium hydroxide + hydrogen

sodium + oxygen   sodium oxide + hydrogen

17.4 Suggest why Group 1 elements are kept in oil. **[1 mark]**

Knowledge

C4 Metals

Metallic structure and properties

The atoms that make up metals form layers. The **electrons** in the outer shells of the atoms are **delocalised** – this means they are free to move through the whole structure. The positive metal ions are then attracted to these delocalised electrons by the electrostatic force of attraction.

Pure metals are malleable (soft) because the layers can slide over each other. Metals are good conductors of electricity and of thermal energy because delocalised electrons are free to move through the whole structure. Metals have high melting and boiling points because the electrostatic force of attraction between metal ions and delocalised electrons is strong so lots of energy is needed to break it.

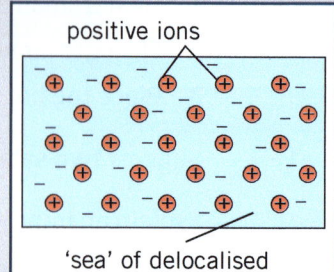

Alloys

Pure metals are often too soft to use as they are. Adding atoms of a different element can make the resulting mixture harder because the new atoms will be a different size to the pure metal's atoms. This will disturb the regular arrangement of the layers, preventing them from sliding over each other.

The harder mixture is called an alloy.

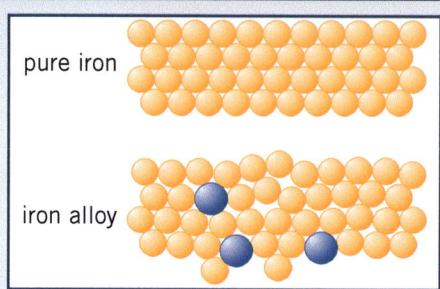

Copper

Copper is useful for electrical wiring and plumbing because it is a good conductor of electricity and heat, can be bent into pipes and tanks, and doesn't react with water.

Reactivity series

The reactivity of a metal is how chemically reactive it is. Some metals react with water very vigorously. Other metals barely react with water or acid. The **reactivity series** places metals in order of their reactivity. Carbon and hydrogen are included for comparison, even though they aren't metals.

Reaction with water	Reaction with acid	Reactivity series		Extraction method
		Metal	Reactivity	
fizzes, gives off hydrogen gas	explodes	potassium	high reactivity	electrolysis
		sodium		
		lithium		
reacts very slowly	fizzes, gives off hydrogen gas	calcium	decreasing reactivity	
		magnesium		
		aluminium		
		(carbon)		
		zinc		reduction with carbon
		iron		
no reaction	reacts slowly with warm acid	tin		
		lead		
		(hydrogen)		
	no reaction	copper	low reactivity	mined from the Earth's crust
		silver		
		gold		

Key Terms

Make sure you can write a definition for these key terms.

delocalised   electron   displacement   electrolysis   extraction   half equation   ion
ionic equation   malleable   metal ore   oxidation   reactivity series   recycling   redox
reduction   spectator ion   thermal decomposition

C4

Displacement reactions

In a **displacement** reaction a *more* reactive element takes the place of a *less* reactive element in a compound. For example, iron is more reactive than copper, so iron displaces the copper in copper sulfate:

copper sulfate + iron → iron sulfate + copper

$CuSO_4(aq) + Fe(s) \rightarrow FeSO_4(aq) + Cu(s)$

Ionic half equations

When an ionic compound is dissolved in a solution, we can write the compound as its separate ions. For example, $CuSO_4(aq)$ can be written as $Cu^{2+}(aq)$ and $SO_4^{2-}(aq)$.

We can write **half equations** to show what happens to the separate ions during a reaction. For example, an iron atom loses two electrons to form an iron ion:

$Fe(s) \rightarrow Fe^{2+}(aq) + 2e^-$

A copper ion gains two electrons to form a copper atom:

$Cu^{2+}(aq) + 2e^- \rightarrow Cu(s)$

Reduction and oxidation

If a substance gains oxygen in a reaction, it has been **oxidised**. If a substance loses oxygen in a reaction, it has been **reduced**. These are called **redox** reactions. They can also be defined as the loss and gain of electrons. Oxidation is the *loss* of electrons, and reduction is the *gain* of electrons. For example:

iron + oxygen → iron oxide

iron has been oxidised

iron oxide + carbon → iron + carbon dioxide

iron oxide has been reduced

Metal carbonates

Metals are commonly found as metal carbonates. Most metal carbonates can undergo **thermal decomposition**. When they are heated, they break up into a metal oxide and water. For example:

copper carbonate → copper oxide + carbon dioxide

Group 1 metal carbonates, apart from lithium, do not thermally decompose.

All metal carbonates react with acids to form a salt, carbon dioxide, and water. For example:

calcium carbonate + hydrochloric acid → calcium chloride + carbon dioxide + water

Metal extraction

Some unreactive metals, like gold, are found as pure metals in the Earth's crust and can be mined. Most metals exist as **ores**, compounds in the rock, and must be extracted. Metals that are less reactive than carbon can be extracted by **reduction** with carbon. Metals that are more reactive than carbon can be extracted using a process called **electrolysis**.

Metal ores are a finite resource and these processes require lots of energy. **Recycling** metals is a good way of reducing the cost and environmental impact. Phytomining and bioleaching are two alternative processes used to extract copper from low-grade ores (ores with only a little copper in them). Both methods avoid moving large amounts of rock needed in traditional mining techniques.

Phytomining

1. Grow plants near the metal ore.
2. Harvest and burn the plants.
3. The ash contains the metal compound.
4. Process the ash by electrolysis or displacement with scrap metal.

Bioleaching

1. Grow bacteria near the metal ore.
2. Bacteria produce leachate solutions that contain metal compound.
3. Process the leachate by electrolysis or displacement with scrap metal.

Retrieval

Learn the answers to the questions below then cover the answers column with a piece of paper and write down as many as you can. Check and repeat.

C4 questions | Answers

#	Question	Answer
1	What is an ion?	atom that has lost or gained electrons
2	How are metals more reactive than carbon extracted?	electrolysis
3	Why is gold found as a native metal in the ground?	it is unreactive
4	What do we call the process of removing oxygen from a compound?	reduction
5	What force holds metal ions and the delocalised electrons together?	the electrostatic force of attraction
6	What do we call a reaction where an atom loses electrons?	oxidation
7	What is reduction, in terms of electrons?	gaining electrons
8	What happens in a displacement reaction?	a more reactive element takes the place of a less reactive one in a compound
9	What are low-grade ores?	ores that are not financially worth extracting by normal means
10	Name two ways of extracting low-grade ores.	phytomining and bioleaching
11	What is used to extract the metals in phytomining?	plants
12	Describe the structure and bonding in a pure metal.	layers of positive metal ions with delocalised electrons, strong electrostatic forces of attraction between metal ions and delocalised electrons
13	Give four properties of pure metals.	malleable, high melting/boiling points, good conductors of electricity, good conductors of thermal energy
14	Explain why pure metals are malleable.	layers can slide over each other easily
15	Explain why metals have high melting and boiling points.	electrostatic force of attraction between positive metal ions and delocalised electrons is strong and requires a lot of energy to break
16	Explain why metals are good conductors of electricity and of thermal energy.	delocalised electrons are free to move through the metal
17	What is an alloy?	mixture of a metal with atoms of another element
18	Explain why alloys are harder than pure metals.	different sized atoms disturb the layers, preventing them from sliding over each other

40 C4 Metals

C4

Now use the questions below to check your knowledge from previous chapters

Previous questions | Answers

#	Question	Answer
1	Which number is found below the atomic symbol on the Periodic Table?	atomic (proton) number
2	How can you find out the number of protons in an atom?	atomic number on Periodic Table
3	How can you find out the number of neutrons in an atom?	mass number – atomic number
4	Why are the numbers of electrons and protons equal in an atom?	to give no overall charge
5	Explain why large molecules have higher melting and boiling points than small molecules.	the intermolecular force is stronger
6	Why do most covalent substances not conduct electricity?	they do not have delocalised electrons
7	Describe the structure and bonding in graphite.	each carbon atom is bonded to three others in hexagonal rings arranged in layers; it has delocalised electrons and no bonds between the layers
8	Explain why graphite can conduct electricity.	the delocalised electrons can move through the graphite
9	Explain why graphite is soft.	there are weak intermolecular forces between the layers that are easily broken so the layers can slide over each other
10	What is graphene?	one layer of graphite
11	Why does the reactivity of the alkali metals increase down the group?	bigger atoms, outer electron further from nucleus, weaker electrostatic force of attraction, easier to lose the electron in the outer shell
12	What name is given to the Group 7 elements?	halogens
13	Give the formulae of the first four halogens.	F_2, Cl_2, Br_2, I_2

Maths Skills

Practise your maths skills using the worked example and practice questions below.

Rearranging equations	Worked example	Practice
You need to be able to rearrange and apply many equations in chemistry, for example, the equation to calculate number of moles. Chemists use moles to describe the relative numbers of particles in a given mass of substance. This can be calculated using: $$\text{number of moles} = \frac{\text{mass (g)}}{\text{relative atomic mass }(A_r)\text{ or relative formula mass }(M_r)}$$ This equation can be rearranged to find the mass of a substance, or the A_r or M_r.	What is the mass of 7.5×10^{-3} moles of aluminium sulfate? **Answer:** Aluminium sulfate = $Al_2(SO_4)_3$ M_r of $Al_2(SO_4)_3 = (27 \times 2) + (32 \times 3) + (16 \times 12) = 342$ Rearrange the equation: mass = number of moles $\times M_r$ mass = $(7.5\times10^{-3}) \times 342 = 2.565 = 2.6\,g$	1 Calculate the relative formula mass of H_2SO_4. 2 Calculate the number of moles of neon atoms in 0.02 g of neon. 3 Calculate the mass of copper sulfate produced through evaporating 1.5 mol copper sulfate solution.

C4 Retrieval 41

Practice

Exam-style questions

01.1 Identify which property is typical of metals.
Tick **one** box. **[1 mark]**

- They are poor conductors of electricity. ☐
- They conduct electricity in the solid state but not in the liquid state. ☐
- They conduct electricity in the liquid state but not in the solid state. ☐
- They conduct electricity in the solid and liquid states. ☐

> **Exam Tip**
> Only tick one box. Ticking more than one will mean no marks, even if one of the boxes you've ticked is correct.

01.2 Describe the structure of a pure metal. **[3 marks]**

_____
_____
_____
_____

01.3 Explain why the bonding in a pure metal means that metals can be shaped. **[2 marks]**

_____
_____
_____

> **Exam Tip**
> This is another way of asking why pure metals are soft.

01.4 Mercury is a metal. It is a liquid at room temperature. Suggest why mercury is an unusual metal. **[1 mark]**

_____
_____

02 This question is about the reactivity series.

02.1 Identify which of these pairs of substances react together in a displacement reaction.
Tick **one** box. **[1 mark]**

- zinc and magnesium chloride solution ☐
- zinc and copper chloride solution ☐
- iron and zinc sulfate solution ☐
- iron and magnesium sulfate solution ☐

> **Exam Tip**
> You are expected to remember the order of elements within the reactivity series. There are lots of rhymes to help you; just pick the one you like the best.

42 C4 Metals

C4

02.2 Name the gas formed when sodium reacts with water. **[1 mark]**

_____

02.3 Lithium also reacts with water. Does lithium react more vigorously or less vigorously with water than sodium?
Draw a circle around **one** answer. **[1 mark]**

     more vigorously   less vigorously

> **! Exam Tip**
>
> Identify the position of lithium and sodium on the Periodic Table to help with this question.

02.4 Lithium is added to a solution of copper chloride. Name the substance formed. **[1 mark]**

_____

02.5 Use your answers to questions **02.1** through **02.4** to put copper, lithium, sodium, and zinc in order of reactivity. **[3 marks]**

most reactive   least reactive

_____ > _____ > _____ > _____

03 Iron is found on Earth as iron(III) oxide. To obtain pure iron, iron(III) oxide is reacted with carbon.

03.1 Identify whether the iron is oxidised or reduced. Give a reason for your answer. **[2 marks]**

03.2 Balance the symbol equation for the extraction of iron from iron(III) oxide. **[1 mark]**

_____ Fe_2O_3 + _____ C → _____ Fe + _____ CO_2

> **! Exam Tip**
>
> Only write numbers in the boxes, don't try to change the formulae of the compounds already given.

03.3 Explain why you do not need to react gold with carbon to obtain pure gold. **[2 marks]**

04 A student pours some dilute hydrochloric acid into a beaker. The student then adds some pieces of zinc to the acid.

04.1 Describe **one** observation the student would make. **[1 mark]**

04.2 Name the **two** products of the reaction. **[1 mark]**

04.3 Write a balanced chemical equation for the reaction, including state symbols. **[3 marks]**

04.4 Explain whether zinc is oxidised or reduced in the reaction. **[2 marks]**

> **! Exam Tip**
>
> An observation is what you would _see_ during this reaction.

> **! Exam Tip**
>
> If you forget the state symbols, you won't get all of the marks.

C4 Practice 43

05 Figure 1 shows three solutions in test tubes.

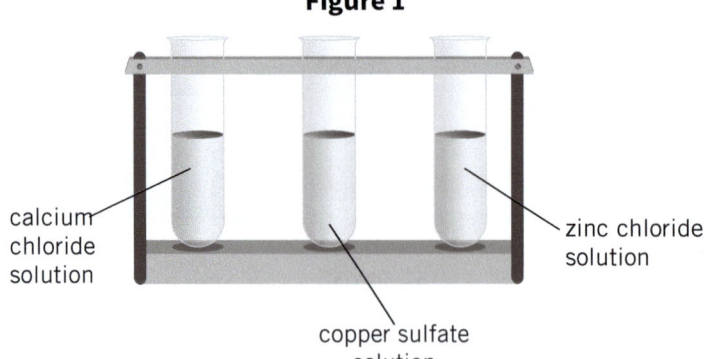

Figure 1

A student added small pieces of metal **X** to each test tube. **Table 1** shows the student's observations.

Table 1

Solution	Observations
calcium chloride	no change
copper sulfate	brown solid forms on the surface of metal **X** and blue colour of solution becomes paler
zinc chloride	grey solid forms on the surface of metal **X**

05.1 Suggest a possible identity of metal **X**. Justify your prediction. **[3 marks]**

> **Exam Tip**
> Only giving a metal will not be enough to get all the marks; you must explain why you picked that one.

05.2 Write a balanced chemical equation for the reaction of metal **X** with copper sulfate solution. Use the symbol **X** to represent the metal, and assume it forms **X^{2+}** ions. Include state symbols in your equation. **[3 marks]**

05.3 Explain how the student could confirm the identify of metal **X**. **[4 marks]**

06 Platinum is a metal. It can be used to make jewellery. **Figure 2** shows the arrangement of particles in platinum.

Figure 2

06.1 Explain why platinum can be bent and shaped. Give your answer in terms of the arrangement of particles in the metal. **[2 marks]**

06.2 Explain why platinum has a high melting point. Give your answer in terms of the bonding in the metal. **[3 marks]**

> **Exam Tip**
> This is a two-mark question so you need to write at least two points.

44 C4 Metals

06.3 Pure platinum is quite soft. An alloy of platinum that contains rhodium is harder than pure platinum. Suggest an advantage of making jewellery from a platinum alloy instead of from pure platinum. **[1 mark]**

> **Exam Tip**
> Think carefully about the difference between the structures of pure platinum and the platinum alloy.

06.4 Explain why the platinum alloy is harder than pure platinum. **[3 marks]**

07 Magnesium reacts with oxygen to form magnesium oxide.

07.1 **Figure 3** shows a model of magnesium atoms. Complete **Figure 3** to show the metallic bonding in magnesium. **[2 marks]**

> **Exam Tip**
> You'll need to add on positive and negative charges.

Figure 3

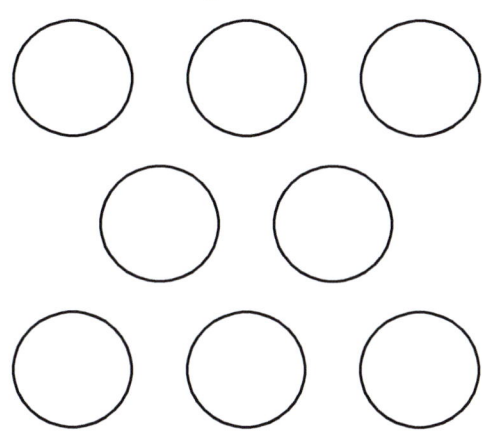

07.2 Draw a dot and cross diagram of magnesium oxide. **[2 marks]**

07.3 Draw **one** line from each substance to the correct property and explanation. **[2 marks]**

> **Exam Tip**
> Read the instruction in the question carefully.
> Not all of the boxes on the right hand side will have lines going to them.

Substance	Property and explanation
	conducts electricity in the solid and liquid states because its electrons are free to move
magnesium oxide	conducts electricity in the liquid state only because its electrons are then free to move
magnesium	conducts electricity in the solid and liquid states because its ions are free to move
	conducts electricity in the liquid state only because its ions are then free to move

07.4 Write a balanced symbol equation for the reaction between magnesium and oxygen. Include state symbols. **[3 marks]**

> **Exam Tip**
> When writing balanced equations:
> Step 1: recall the formulae
> Step 2: write down the reactants
> Step 3: work out the ions
> Step 4: determine the formulae of any products
> Step 5: check the equation is balanced
> Step 6: add state symbols

07.5 Magnesium and aluminium are two metals. They both have a low density, which makes them lightweight metals. Bicycle wheels can be made of magnesium and aluminium alloys. Explain why an alloy of magnesium and aluminium is used instead of the pure metals. **[2 marks]**

08 **Table 2** shows the relative conductivities of some metals. The higher the relative conductivity value, the better the metal conducts electricity.

Table 2

Metal	Relative conductivity
aluminium	0.382
beryllium	0.250
lithium	0.108
magnesium	0.224
sodium	0.218
zinc	0.167

08.1 Describe the bonding in pure metals. **[2 marks]**

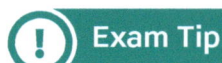

08.2 "The conductivity of a metal depends on the number of delocalised electrons per atom and which period the metal is in."

Zinc has two electrons in its outer shell. Evaluate the statement above using **Table 2** and the Periodic Table. **[6 marks]**

You'll have to give evidence for and against this statement and then state a conclusion and justify it with data from the table.

09 **Table 3** shows the properties of five substances. Each substance is represented by a letter. The letters are not the chemical formulae of the substances.

Table 3

Substance	Melting point in °C	Does it conduct electricity in the solid state?	Does it conduct electricity in the liquid state?
A	993	no	yes
B	1085	yes	yes
C	1263	no	yes
D	1064	yes	yes
E	30	yes	yes

09.1 Use data from **Table 3** to deduce whether substance **C** is a metal or an ionic compound. Justify your answer. **[3 marks]**

09.2 Give the letters of **two** substances in **Table 3** that could be the element copper. Justify your answer. **[2 marks]**

Before you start, mark which compound is ionic and which is covalent, this will prevent you from getting confused later.

09.3 Substance **E** represents the metal gallium. Explain why the melting point of gallium is unusual compared to most other metals.

[3 marks]

C4 Metals

10 Table 4 shows some data about four substances.

Table 4

Substance	State at 25 °C	Melting point in °C	Boiling point in °C	Conducts electricity when solid	Conducts electricity when liquid
A		−219	−183	no	no
B	solid	1538	2862	yes	yes
C	solid	801	1465	no	yes
D		−7	59	no	no

10.1 Complete **Table 4** to show the state of substance **A** and substance **D**. **[2 marks]**

10.2 Explain how you can tell that substance **B** is a metal. **[2 marks]**

> **Exam Tip**
> Always use evidence from the data given to you.

10.3 Identify which substance is sodium chloride. Explain your answer. **[4 marks]**

10.4 Identify which substance in **Table 4** represents oxygen. **[1 mark]**

10.5 Describe the structure and bonding in substance **D**. **[2 marks]**

11 A student is investigating metal displacement reactions. The student places small pieces of metal in the depressions of a 3 × 3 white spotting tile. The student then adds small amounts of solutions of metal salts to the metals. **Table 5** shows the metals and solutions of metal salts that the student used.

Table 5

magnesium + magnesium chloride solution	zinc + magnesium chloride solution	copper + magnesium chloride solution
magnesium + zinc chloride solution	zinc + zinc chloride solution	copper + zinc chloride solution
magnesium + copper chloride solution	zinc + copper chloride solution	copper + copper chloride solution

11.1 Suggest **one** improvement to the experiment that would prevent the unnecessary use of metals and solutions. **[1 mark]**

11.2 Name the metal that does **not** react with any of the solutions. Explain your choice. **[2 marks]**

> **Exam Tip**
> Find the metals on the reactivity series.

11.3 Write an ionic equation with state symbols for the reaction of magnesium with copper(II) sulfate solution. **[3 marks]**

11.4 Write the electronic structure of magnesium before and after its reaction with copper(II) sulfate solution. **[2 marks]**

> **Exam Tip**
> This isn't drawing it out, but writing the number of electrons. It starts with 2,8...

11.5 Suggest **three** advantages of doing this experiment on a spotting tile compared to using test tubes.
Give a reason for each suggestion. **[3 marks]**

12 Copper exists naturally on the Earth chemically bonded to non-metals. The pure metal can be extracted from these compounds.

12.1 Describe how copper is obtained from low-grade copper ore by phytomining. **[4 marks]**

Exam Tip

For an evaluate question you need four parts:
- the good
- the bad
- your opinion
- the reason for your opinion

12.2 Evaluate the advantages and disadvantages of obtaining copper by recycling scrap copper and bioleaching. **[6 marks]**

12.3 An ore of copper contains 22.1% copper. Calculate the mass of waste produced when 50.0 kg of copper is extracted from the ore. **[4 marks]**

13 Most metals are used as alloys.

13.1 Explain why pure metals are alloyed. **[4 marks]**

13.2 Complete the sentences using answers from the box. Each word can be used once, more than once, or not at all. **[4 marks]**

| carbon | copper | gold | magnesium | tin | zinc |

Bronze is an alloy that is made from _____ and _____. Brass is an alloy made from _____ and _____.

13.3 Gold is often used to make jewellery. Suggest why jewellery made from gold is more valuable than jewellery made from similar coloured metals like copper. **[1 mark]**

13.4 The purity of a sample of gold is measured in carats. 24 carat gold is pure gold. 18 carat gold is an alloy made of 75% gold mixed with other materials. Calculate the percentage of gold in a 14 carat gold sample. **[2 marks]**

Exam Tip

You probably haven't come across a question in this exact context but you need to get used to applying the maths you know in new situations.

14 In a decomposition reaction, one substance breaks down on heating to form two or more substances.

14.1 Magnesium nitrate decomposes to form solid magnesium oxide, nitrogen dioxide gas, and oxygen gas. Write a balanced chemical equation for the reaction, including state symbols. **[3 marks]**

14.2 The equation for another decomposition reaction is:

$$CaCO_3(s) \rightarrow CaO(s) + CO_2(g)$$

Give the name of the compound with the formula $CaCO_3$. **[1 mark]**

14.3 Calculate the relative formula mass of $CaCO_3$. **[2 marks]**

14.4 50 kg of $CaCO_3$ is heated until the decomposition reaction is complete. The reaction produces 22 kg of CO_2. Calculate the mass of CaO that is produced. **[2 marks]**

Exam Tip

Don't forget the state symbols – they have a mark associated with them, so remember to include them.

C4 Metals

C4

15 This question is about the production of the metal lead. Methods **A** and **B** describe two processes for extracting lead.

Method A
1. Dig lead sulfide from the ground.
2. Heat the lead sulfide in air: $2PbS + 3O_2 \rightarrow 2PbO + 2SO_2$
3. Heat the lead oxide with carbon: $2PbO + C \rightarrow 2Pb + CO_2$

Method B
1. Collect battery paste from used batteries. The paste is a mixture of lead sulfate and lead oxides.
2. React the paste with an alkaline solution. One product is a soluble sulfate solution.
3. Heat the remaining mixture with carbon, for example:
$2PbO + C \rightarrow 2Pb + CO_2$

> **Exam Tip**
>
> Remember OILRIG
> **O**xidation
> **I**s
> **L**oss (of electrons)
> **R**eduction
> **I**s
> **G**ain (of electrons)

15.1 Name the substance that is reduced in step **3** of method **A**. **[1 mark]**

15.2 Write a balanced chemical equation for step **2** of method **B**.
- The reactants are lead sulfate and sodium hydroxide.
- The products are lead hydroxide and sodium sulfate. **[3 marks]**

15.3 Suggest **two** advantages of method **B** compared to method **A**. **[2 marks]**

16 Some metals react with water.

16.1 Describe the expected observations in the reaction of potassium and water. **[2 marks]**

16.2 Name the products of the reaction of lithium with water. **[2 marks]**

16.3 Describe the expected observations when copper is placed in a test tube of water. **[1 mark]**

> **Exam Tip**
>
> An observation is what you *see* happening.

17 Carbon forms many different structures.

17.1 Complete the sentences. **[3 marks]**

The carbon in diamond makes _____ carbon–carbon bonds.

Diamond molecules are very large and has a very _____ melting point.

Diamond is very _____.

17.2 Explain why graphite has similar properties to metals. **[2 marks]**

> **Exam Tip**
>
> This is a two-mark explain question. This means that your answer should include a why.

Knowledge

C5 Electrolysis

Electrolysis

In the process of **electrolysis**, an electric current is passed through an **electrolyte**. An electrolyte is a liquid or solution that contains ions and so can conduct electricity. This causes the ions to move to the **electrodes**, where they form pure elements.

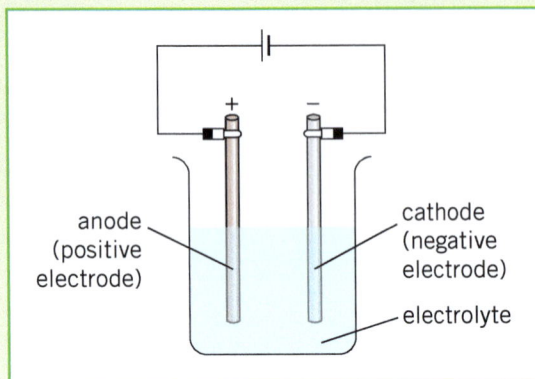

Electrolysis of aqueous solutions

Solid ionic compounds can also undergo electrolysis when dissolved in water. It requires less energy to dissolve ionic compounds in water than it does to melt them. However, in the electrolysis of solutions, the pure elements are not always produced. This is because the water can also undergo ionisation:

$$H_2O(l) \rightarrow H^+(aq) + OH^-(aq)$$

Products at the anode

In the electrolysis of a solution, if the non-metal contains oxygen then oxygen gas is formed at the anode:

- The $OH^-(aq)$ ions formed from the ionisation of water are attracted to the anode.
- The $OH^-(aq)$ ions lose electrons to the anode and form oxygen gas.
- $4OH^-(aq) \rightarrow O_2(g) + 2H_2O(l) + 4e^-$

If the non-metal ion is a halogen in high concentration, then the halogen gas is formed at the anode:

- $2Cl^-(aq) \rightarrow Cl_2(g) + 2e^-$

Electrolysis of molten compounds

Solid ionic compounds do not conduct electricity as the ions cannot move. To undergo electrolysis they must be molten or dissolved, so the ions are free to move.

When an ionic compound is molten:

- the positive metal ions are *attracted* to the **cathode**, where they will *gain* electrons to form the pure metal (**reduction**)
- the negative non-metal ions are *attracted* to the **anode**, where they will *lose* electrons and become the pure non-metal (**oxidation**).

For example, molten sodium chloride, NaCl, can undergo electrolysis to form sodium at the cathode and chlorine at the anode.

Half equations

sodium chloride → sodium + chlorine

$$2NaCl(l) \rightarrow 2Na(s) + Cl_2(g)$$

- at the cathode: $2Na^+(l) + 2e^- \rightarrow 2Na(s)$
- at the anode: $2Cl^-(l) \rightarrow Cl_2(g) + 2e^-$

Products at the cathode

In the electrolysis of a solution, if the metal is more **reactive** than hydrogen then hydrogen gas is formed at the cathode:

- The $H^+(aq)$ ions from the ionisation of water are attracted to the cathode and react with it.
- The $H^+(aq)$ ions gain electrons from the cathode and form hydrogen gas.
- $2H^+(aq) + 2e^- \rightarrow H_2(g)$
- The metal ions remain in solution.

Electrolysis of sodium chloride solution

When sodium chloride solution (brine) is electrolysed, it makes three commercially valuable products:

- hydrogen used in many chemical processes
- chlorine used to make bleach and plastic
- sodium hydroxide used to make soap

Reactivity series (most reactive → least reactive):
potassium, sodium, calcium, magnesium, aluminium, **(carbon)**, zinc, iron, tin, lead, **(hydrogen)**, copper, silver, gold, platinum

50 C5 Electrolysis

C5

Electroplating

Electroplating uses electrolysis to coat one metal with a thin layer of another, more precious and less reactive metal. This makes the object more desirable, more durable, or protects it from corrosion. For example, cheap jewellery can be silver-plated using electroplating.

Electrolysis of aluminium oxide

Electrolysis can be used to extract metals from their ionic compounds. Electrolysis is used if the metal is more reactive than carbon. Aluminium is extracted from aluminium oxide by electrolysis:

1. The aluminium oxide is mixed with a substance called **cryolite**, which lowers the melting point.
2. The mixture is then heated until it is molten.
3. The resulting molten mixture undergoes electrolysis.

aluminium oxide → aluminium + oxygen

$2Al_2O_3(l) \rightarrow 4Al(l) + 3O_2(g)$

cathode: pure aluminium is formed $Al^{3+}(l) + 3e^- \rightarrow Al(l)$

anode: oxygen is formed $2O^{2-}(l) \rightarrow O_2(g) + 4e^-$

In the electrolysis of aluminium, the anode is made of graphite. The graphite reacts with the oxygen to form carbon dioxide and so slowly wears away. It therefore needs to be replaced frequently.

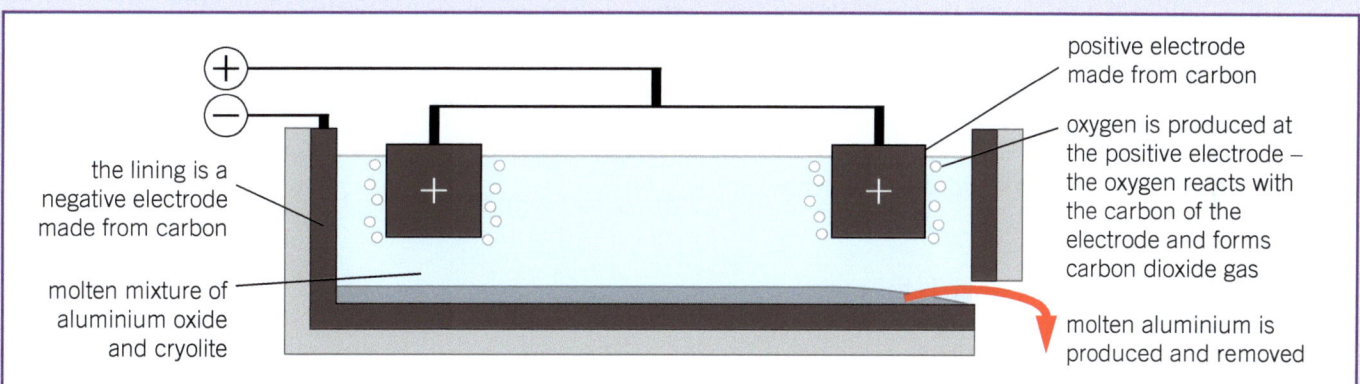

> **Revision Tip**
>
> Extraction of aluminium by electrolysis makes a good six-mark question. It's a bit different from the other example of electrolysis that you need to learn. Make sure you can do the half equations, know the key words, and know what the electrodes are made of.

> **Revision Tip**
>
> In an exam, don't PANIC.
> Here is an easy way to remember which electrode is which:
> **P**ositive
> **A**node
> **N**egative
> **I**s
> **C**athode

 **Key Terms** — Make sure you can write a definition for these key terms.

| anode | cathode | cryolite | electrode | electrolysis |
| electrolyte | electroplating | oxidation | reactivity | reduction |

C5 Knowledge   51

Retrieval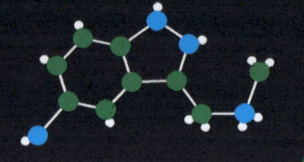

Learn the answers to the questions below then cover the answers column with a piece of paper and write as many as you can. Check and repeat.

C5 questions | Answers

#	Question	Answer
1	What is electrolysis?	process of using electricity to extract elements from a compound
2	What is an electrode?	the end of a circuit which is placed in the electrolyte
3	What is an electrolyte?	the liquid or solution that electrolysis is carried out in
4	What is the cathode?	the negative electrode
5	What is the anode?	the positive electrode
6	Where are metals formed?	at the cathode
7	Where are non-metals formed?	at the anode
8	How can ionic substances be electrolysed?	by melting or dissolving them
9	Why can solid ionic substances not be electrolysed?	they do not conduct electricity
10	In the electrolysis of aluminium oxide, why is the aluminium oxide mixed with cryolite?	to lower the melting point
11	In the electrolysis of aluminium oxide, why do the anodes need to be replaced?	they react with the oxygen being formed
12	In the electrolysis of solutions, when is the metal not produced at the cathode?	when the metal is more reactive than hydrogen
13	In the electrolysis of solutions, what is produced at the anode?	a halogen or oxygen
14	What are the three products of the electrolysis of sodium chloride solution?	hydrogen, sodium hydroxide, chlorine
15	What are the reasons for electroplating a metal?	increase durability, improve desirability, reduce corrosion
16	Where does oxidation happen in electrolysis?	at the cathode
17	Where does reduction happen in electrolysis?	at the anode
18	What carries the charge through the electrolyte?	the ions that can move

C5 Electrolysis

C5

Now use the questions below to check your knowledge from previous chapters.

Previous questions | Answers

#	Question	Answer
1	What is the relative mass of a proton?	1
2	What is the relative mass of a neutron?	1
3	What is the relative mass of an electron?	0 (very small)
4	How are covalent bonds formed?	atoms sharing electrons
5	How many electrons go into a covalent bond?	2 for a single bond, 4 for a double bond
6	Between which kinds of atom does covalent bonding occur?	non-metals
7	What are the two main types of covalent structure?	giant covalent, small molecules
8	Describe the structure and bonding of a giant covalent substance.	billions of atoms bonded together with strong covalent bonds
9	What is an ion?	an atom that has lost or gained electrons

(Put paper here)

Required Practical Skills

Practise answering questions on the required practicals using the example below.
You need to be able to apply your skills and knowledge to other practicals too.

Electrolysis	Worked example	Practice
You need to be able to describe the method of electrolysis, and label the experimental set-up for electrolysis. Electrolysis uses electricity to break ionic compounds down into simpler compounds or elements. Metals or hydrogen are made at the negative electrode, and non-metal molecules are made at the positive electrode. You will need to be able to apply the principles of electrolysis to any example, as many solutions can undergo electrolysis. This includes predicting the products of electrolysis for different solutions, identifying which ions move to each electrode, and writing equations for the reactions at the two electrodes.	The electrolysis of aqueous copper sulfate gives two products. Identify these products and the electrodes they form at. State three observations you would make. **Answer:** The two products are copper (Cu) and oxygen gas (O_2). The copper forms at the negative electrode and the oxygen gas forms at the positive electrode. The copper will be an orange metal at the negative electrode. The oxygen will be seen as bubbles at the positive electrode. The bubbles would reignite a glowing splint. Over time, the blue colour of copper sulfate will disappear.	1 State what you would observe at each electrode during the electrolysis of copper(II) chloride. 2 Give the products of the electrolysis of sodium sulfate. 3 Explain why the electrodes must not touch each other during electrolysis.

C5 Retrieval

Practice

Exam-style questions

01 A student investigated the electrolysis of sodium chloride solution.

Figure 1 shows the apparatus used.

Figure 1

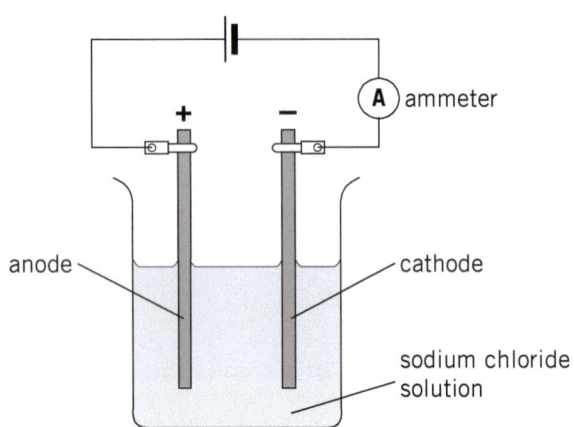

01.1 Name the substance that is produced at the cathode. **[1 mark]**

_____

> **! Exam Tip**
> It's not sodium. Use the reactivity series and the formula of salty water to work out the other product.

01.2 Write a half equation, including state symbols, for the reaction that occurs at the anode. **[3 marks]**

_____

01.3 The student wanted to investigate if changing the concentration of sodium chloride solution affects the current that flows.

Table 1 shows the student's results.

Table 1

Concentration of sodium chloride solution in mol/dm³	Current in amps
0.2	0.20
0.4	0.33
0.6	0.43
0.8	0.47
1.0	0.52

Identify the independent variable and the dependent variable in the investigation. **[2 marks]**

independent variable: _____

dependent variable: _____

01.4 Suggest **one** control variable in the investigation. **[1 mark]**

_____

> **! Exam Tip**
> 'Independent' is the one we change and 'dependent' is the one we measure. A good way to remember this is that your results *depend* on the dependent variable.

54   C5 Electrolysis

C5

01.5 Plot the data from **Table 1** on **Figure 2**.
Draw a line of best fit. **[3 marks]**

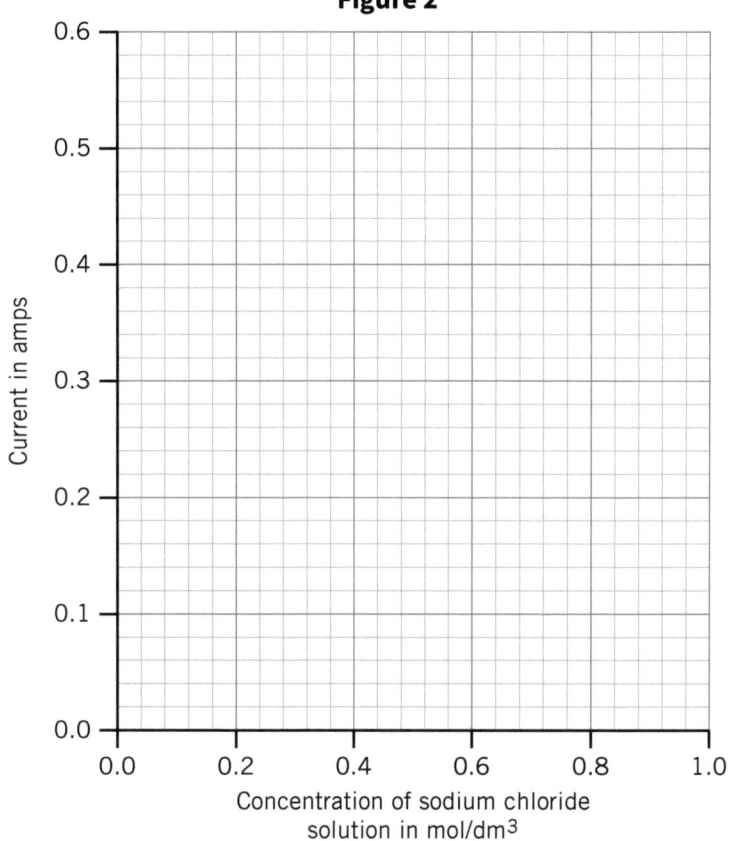

!) **Exam Tip**

Use crosses to plot your points because this clearly shows the examiners which point you are aiming for. Circles can easily be misinterpreted as they can cover a range of points or be too small to be seen by the examiner. Crosses are the best way to ensure you get the mark.

01.6 Describe the pattern shown on your graph.
Suggest **one** reason for this pattern. **[2 marks]**

pattern: _____

_____

reason: _____

_____

!) **Exam Tip**

Use data from the graph to support your reason.

02 A teacher passed an electric current through molten zinc chloride. **Figure 3** shows the apparatus.

Figure 3

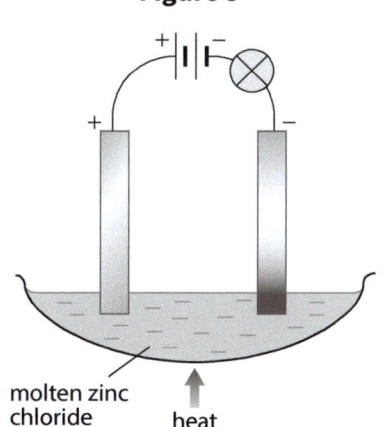

C5 Practice 55

02.1 Predict the observations at the positive and negative electrodes. **[2 marks]**

_____

_____

02.2 Write a half equation for the reaction that occurs at the negative electrode. **[3 marks]**

_____

_____

> **! Exam Tip**
> Start by working out the ions that will be in zinc chloride and to which electrode they will be attracted.

02.3 The teacher passes an electric current through an aqueous zinc chloride solution. Predict the products formed at the positive and negative electrodes. Zinc is more reactive than hydrogen. **[2 marks]**

03 Molten zinc chloride is electrolysed using inert electrodes.

03.1 Name the electrode that positively charged ions move towards during electrolysis. **[1 mark]**

03.2 What are the products at the anode and cathode? Choose **one** answer. **[1 mark]**

Anode	Cathode
zinc	chlorine
chlorine	zinc
zinc	hydrogen
chlorine	hydrogen

> **! Exam Tip**
> The first step is to work out the charges on the ions within zinc chloride.

03.3 Explain why solid zinc chloride cannot be used for electrolysis. **[3 marks]**

03.4 The symbol equation for the reaction is:

$$ZnCl_2 \rightarrow Zn + Cl_2$$

Complete the symbol equation by adding state symbols. **[1 mark]**

04 Potassium is extracted from its ores by electrolysis.

04.1 Suggest why electrolysis is used to extract potassium. **[1 mark]**

04.2 In the electrolysis of molten potassium sulfate, name the electrode that solid potassium metal will form at. **[1 mark]**

04.3 The electrolysis of molten potassium sulfate is an expensive industrial process. Give a reason why. **[1 mark]**

C5 Electrolysis

04.4 Aqueous potassium sulfate solution can also be electrolysed. Write the half equations for the electrolysis of aqueous potassium sulfate. **[6 marks]**

04.5 Suggest why an aqueous solution of potassium sulfate cannot be used to extract potassium. **[1 mark]**

> **Exam Tip**
> Look at the reactivity series to help you answer this.

05 **Figure 4** shows an electrolysis cell for the industrial extraction of aluminium.

Figure 4

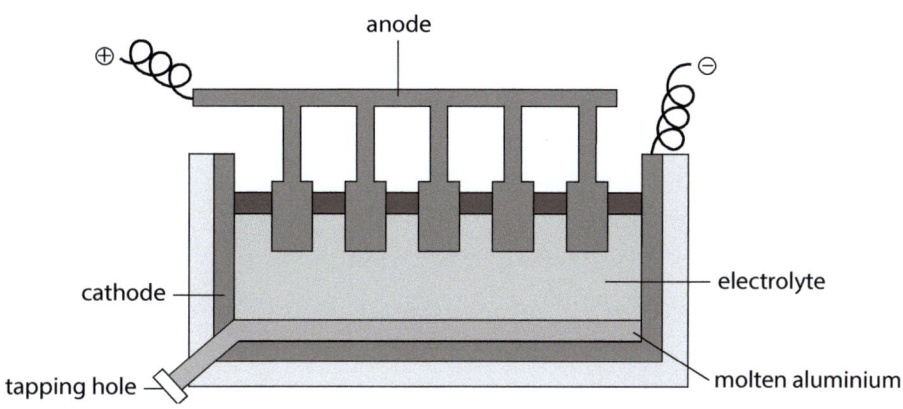

05.1 Explain why aluminium cannot be extracted by heating its ore with carbon. **[1 mark]**

05.2 Name the material that the anode and cathode are made from. **[1 mark]**

05.3 Explain why the anode must be replaced regularly. **[1 mark]**

> **Exam Tip**
> Think about the element that the electrodes are made from.

05.4 Name the **two** substances that are mixed together in the electrolyte. **[2 marks]**

05.5 Write a half equation, including state symbols, for the reaction that occurs at the cathode. **[3 marks]**

> **Exam Tip**
> Make sure the number of electrons matches the charge on the ions, and that the equation is balanced.

05.6 Suggest why industrial aluminium electrolysis cells are often sited near power stations that generate electricity from renewable sources. **[1 mark]**

06.1 Define the term inert electrode. **[2 marks]**

06.2 The charge on a lead ion is 2+. Deduce the formula of lead bromide. **[1 mark]**

06.3 An electrolysis reaction happens when electricity is passed through molten lead bromide using inert electrodes. Describe what happens in this reaction. Include in your answer the name of the products of the electrolysis reaction and an explanation of how the products are made. **[6 marks]**

> **Exam Tip**
> Split your answer into two parts: what happens at the anode and then what happens at the cathode.

07 A student investigates electrolysis cells. **Figure 5** shows the apparatus used.

Figure 5

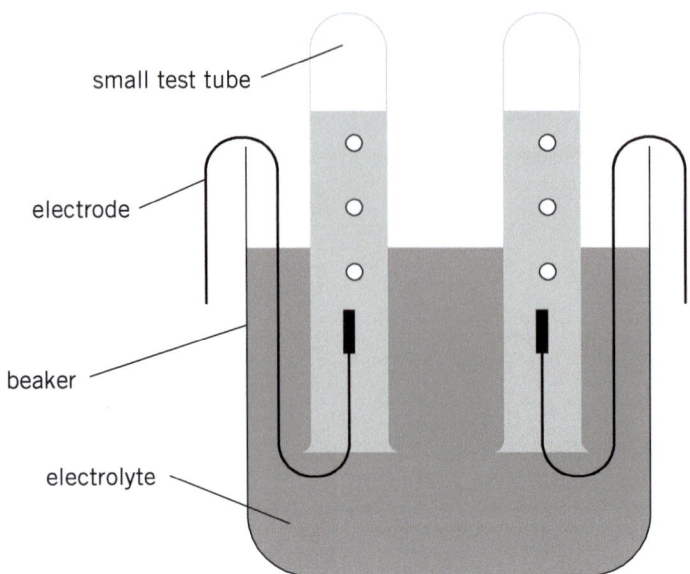

The student used the following method:
1 Set up the electrolysis cell as shown, with a battery to supply the current.
2 Pour the first electrolyte into the beaker.
3 Switch on the current and record any observations.

07.1 Suggest a reason for investigating the electrolysis of aqueous solutions of salts instead of the electrolysis of molten salts. **[1 mark]**

> **Exam Tip**
> Look at what is happening at the electrodes in the diagram.

07.2 Explain why test tubes are placed over the electrodes. **[1 mark]**

07.3 **Table 2** shows some of the student's results.

Table 2

Experiment	Electrolyte	Observations at anode	Observations at cathode
1	copper chloride solution	bubbles	
2	copper sulfate solution	bubbles	cathode coated in reddish metal
3	potassium bromide solution	yellow-brown liquid	bubbles

> **Exam Tip**
> Base your predictions on the other results in the table.

Predict what the student would observe at the cathode in experiment **1**. **[1 mark]**

07.4 Explain how the gas in the bubbles in experiment **3** is formed. Include a half equation in your answer. **[5 marks]**

C5 Electrolysis

08 Aluminium is manufactured by electrolysis.

08.1 Suggest why reduction with carbon is not an appropriate method to manufacture aluminium. **[1 mark]**

08.2 In the electrolysis of aluminium, what is the cathode made of? **[1 mark]**

08.3 A mixture of aluminium oxide and cryolite forms the electrolyte. Explain the purpose of the cryolite. **[3 marks]**

08.4 Explain why aluminium is produced at the cathode. **[2 marks]**

> **Exam Tip**
> The electrodes are made of carbon.

08.5 In the electrolysis of aluminium oxide, explain why the anode has to be replaced regularly. **[2 marks]**

09 A student sets up an electrolysis experiment in a Petri dish, as shown in **Figure 6**.

Figure 6

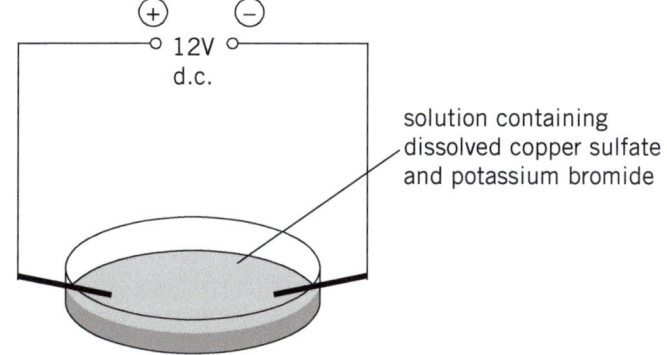

Table 3 shows the results.

Table 3

Electrode	Observations
anode	
cathode	brown flaky solid and then bubbles

09.1 Give the name of the brown flaky solid. **[1 mark]**

09.2 Predict the name of the gas that forms bubbles at the cathode. **[1 mark]**

> **Exam Tip**
> Use the diagram to determine the ions in the electrolyte.

09.3 Describe a test you could do to show that your prediction in **10.2** is correct. Include the expected results of this test. **[2 marks]**

09.4 Predict and explain the observations expected at the anode. Include half equations in your answer. **[8 marks]**

09.5 Suggest **two** reasons for carrying out the electrolysis in a Petri dish, rather than in a larger and taller electrolysis cell. **[2 marks]**

10 Table 4 gives the diameters of some particles.

Table 4

Particle	Diameter in nm
gold atom	0.174
water molecule	0.275

10.1 Explain why a water molecule is **not** a nanoparticle. [1 mark]

10.2 Write the diameter of a water molecule in metres. Give your answer in standard form. [2 marks]

10.3 A certain gold nanoparticle has a cubic shape. The length of a side of the cube is 50 nm. Estimate the number of gold atoms that are on one face of the cube. Give your answer to one significant figure in standard form. [4 marks]

11 This question is about the elements in Group 1 and Group 7 of the Periodic Table.

11.1 Describe the pattern in the melting points of the Group 7 elements, from the top to the bottom of the group. [1 mark]

11.2 Compare the patterns in the reactivity of the Group 1 and Group 7 elements, from the top to the bottom of the groups. [2 marks]

Make sure it's clear which group you're talking about in each part of your answer.

11.3 Name the products formed when sodium reacts with water. [2 marks]

12 The elements calcium and oxygen react together to form an ionic compound called calcium oxide. Use the Periodic Table to help you answer this question.

12.1 Deduce the charge on a calcium ion and write its formula. [1 mark]

12.2 Deduce the charge on an oxide ion and write its formula. [1 mark]

12.3 Predict **three** properties of calcium oxide. Explain why calcium oxide has each of these properties. [6 marks]

Exam Tip

Look at the groups that calcium and oxygen are in and determine the number of electrons in their outer shells. This will tell you how many electrons they lose or gain and then you can work out the charge.

13 Use the Periodic Table to answer the following questions.

13.1 Identify which element is most likely to be used as a catalyst. Choose **one** answer. [1 mark]

calcium   rhodium   sodium   strontium

13.2 Identify which element will react with cold water. Choose **one** answer. [1 mark]

copper   lithium   titanium   zinc

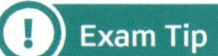

Exam Tip

Only select the number of answers you're asked for in multiple choice questions: in this case, one. If you circle two answers, you won't get the marks. Equally, don't leave any blank!

C5 Electrolysis

13.3 Identify which metal will have the highest density. Choose **one** answer. [1 mark]

aluminium   iron   magnesium   potassium

13.4 Identify which element can form +4 ions. Choose **one** answer. [1 mark]

neon   calcium   sodium   vanadium

13.5 A student has a blue metal compound. Identify which metal the compound contains. Choose **one** answer. [1 mark]

copper   lead   rubidium   tin

14 A student carries out the electrolysis of aqueous copper chloride and sodium chloride.

14.1 Identify the type of bonding in copper chloride and sodium chloride. [1 mark]

14.2 Both copper chloride and sodium chloride produce bubbles when undergoing electrolysis. Identify the electrode that the bubbles will appear at. Choose **one** answer. [1 mark]

anode   cathode

14.3 Name the gas produced in the electrolysis of aqueous copper chloride. [1 mark]

> **! Exam Tip**
> Chloride ions are negatively charged. They will be attracted to the positively charged electrode.

15 A chemist has three magnesium salts (**Table 5**).

Table 5

Magnesium salt	Chemical formula	Melting point in °C
magnesium chloride		714.0
magnesium nitrate	$Mg(NO_3)_2$	88.9
magnesium phosphate	$Mg_3(PO_4)_2$	1184.0

15.1 Complete **Table 5** by writing the chemical formula for magnesium chloride. [1 mark]

15.2 The chemist wants to extract magnesium metal from one of the salts using electrolysis. Identify which salt is the most suitable. Explain your answer. [6 marks]

15.3 Name the electrode at which the magnesium metal is produced. [1 mark]

> **! Exam Tip**
> **Table 5** includes the melting points for the salts, suggesting this is important to consider in your answer.

Knowledge

C6 Chemical analysis

Pure and impure

In chemistry, a **pure** substance contains a single element or compound that is not mixed with any other substance. Pure substances melt and boil at specific temperatures. An **impure** substance contains more than one type of element of compound in a mixture. Impure substances melt and boil at a range of temperatures.

Mixtures

- A mixture consists of two or more elements or compounds that are not chemically combined together.
- The substances in a mixture can be separated using physical processes.
- These processes do not use chemical reactions.

Separating mixtures
- filtration – insoluble solids and a liquid
- crystallisation – soluble solid from a solution
- simple distillation – **solvent** from a solution
- fractional distillation – two liquids with similar boiling points
- paper chromatography – identify substances from a mixture in solution

Chromatography

Chromatography is a method to separate different components in a mixture, for example, foodstuffs and drugs. It is set up as shown here, with a piece of paper in a beaker containing a small amount of solvent.

The R_f **value** is a ratio of how far up the paper a certain spot moves compared to how far the solvent has travelled.

$$R_f = \frac{\text{distance moved by substance}}{\text{distance moved by solvent}}$$

It will always be a number between 0 and 1.

The R_f value depends on the solvent and the temperature, and different substances will have different R_f values. The R_f values for particular solvents can be used to identify a substance.

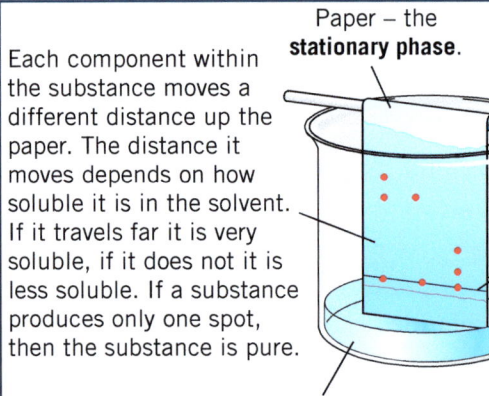

Each component within the substance moves a different distance up the paper. The distance it moves depends on how soluble it is in the solvent. If it travels far it is very soluble, if it does not it is less soluble. If a substance produces only one spot, then the substance is pure.

Paper – the **stationary phase**.

Solvent – the **mobile phase**. The top of the solvent must be below the pencil line or the substances to be tested will dissolve away from the paper.

Solvent front – the top of where the solvent travels up the paper.

The substances to be tested are placed on the pencil line. We draw the line in pencil because pen would dissolve and travel up the paper.

Testing gases

You can identify common gases using the follow tests:

Gas	What you do	What you observe if gas is present
hydrogen	hold a lighted splint near the gas	hear a squeaky pop
oxygen	hold a glowing splint near the gas	splint re-lights
carbon dioxide	bubble the gas through limewater	the limewater turns milky (cloudy white)
chlorine	hold a piece of damp litmus near the gas	a characteristic sharp choking smell, turns blue litmus paper red and then bleaches it white
ammonia	hold a piece of damp blue litmus near the gas hold concentrated hydrochloric acid near the gas	a characteristic sharp choking smell, turns red litmus paper blue, white smoke produced

C6

Testing for cations

Cation	Positive result
aluminium ions, Al^{3+}	on slow addition of excess sodium hydroxide solution, white **precipitate** forms that eventually dissolves again with excess sodium hydroxide
calcium ions, Ca^{2+}	on addition of excess sodium hydroxide solution, white precipitate that does not dissolve
magnesium ions, Mg^{2+}	on addition of excess sodium hydroxide solution, white precipitate that does not dissolve
copper(II) ions, Cu^{2+}	add sodium hydroxide solution, forms a blue precipitate
iron(II) ions, Fe^{2+}	add sodium hydroxide solution, forms a green precipitate
iron(III) ions, Fe^{3+}	add sodium hydroxide solution, forms a brown precipitate

Testing for anions

Anion	Test	Positive result
carbonate, CO_3^{2-}	add dilute acid	carbon dioxide gas formed, which can be tested for with limewater
chloride, Cl^-	add silver nitrate solution in the presence of nitric acid	white precipitate formed
bromide, Br^-	add silver nitrate solution in the presence of nitric acid	cream precipitate formed
iodide, I^-	add silver nitrate solution in the presence of nitric acid	yellow precipitate formed
sulfate, SO_4^{2-}	add barium chloride solution in the presence of hydrochloric acid	white precipitate formed

Flame tests

Substances containing metals can produce a coloured light in a flame. This can be used to identify the metal. However, if there is more than one metal in the substance then this method will not work, as the colours mix. To complete a flame test:

1. Dip a clean nichrome wire loop into a solid sample of the compound.
2. Put the loop into the edge of the blue flame of a Bunsen burner.
3. Observe and record the flame colour produced.

Metal	Flame colour
lithium	crimson
sodium	yellow
potassium	lilac
calcium	orange-red
barium	green

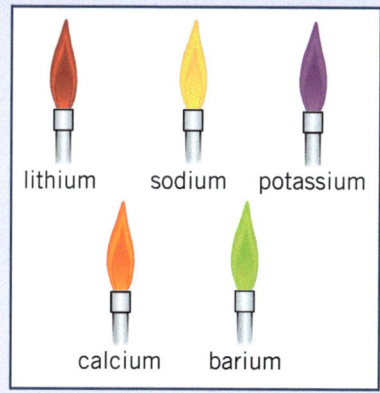

lithium   sodium   potassium

calcium   barium

Key Terms

Make sure you can write a definition for these key terms.

- chromatography
- flame test
- impure
- mobile phase
- precipitate
- pure
- R_f value
- solvent
- solvent front
- stationary phase

Revision Tip

A common exam question is the identification of positive and negative ions in a mystery set of solutions. Make sure you know these tests really well and be prepared to combine them in an exam question.

C6 Knowledge   63

Retrieval

Learn the answers to the questions below then cover the answers column with a piece of paper and write down as many as you can. Check and repeat.

C6 questions | Answers

1. In chemistry, what is a pure substance? — something made of only one type of substance
2. What is chromatography? — a process for separating mixtures
3. What is the stationary phase? — what the mixtures are placed on (usually paper)
4. What is the mobile phase? — the solvent
5. Why is the line drawn in pencil? — ink would dissolve
6. Why must the line be above the solvent? — to stop the substances dissolving into the solvent
7. What does one spot in chromatography mean? — the substance is pure
8. What is the solvent front? — the point which the solvent reaches up to on the stationary phase
9. How is R_f calculated? — $R_f = \dfrac{\text{distance moved by substance}}{\text{distance moved by solvent}}$
10. What is the test for hydrogen? — squeaky pop
11. What is the test for carbon dioxide? — turns limewater milky if bubbled through it
12. What is the test for chlorine? — bleaches damp litmus paper
13. What is the test for aluminium, calcium, and magnesium ions? — forms white precipitate with sodium hydroxide solution
14. How can aluminium ions be distinguished from calcium and magnesium ones? — the white precipitate will dissolve with excess sodium hydroxide
15. What colour precipitates are formed when sodium hydroxide solution is added to solutions of copper(II), iron(II), and iron(III) ions? — copper(II) ions form blue precipitate, iron(II) ions form green precipitate, iron(III) ions form brown precipitate
16. What is the test for a halide ion? — add silver nitrate and nitric acid: chloride forms white precipitate, bromide forms cream precipitate, iodide forms yellow precipitate
17. What is the test for a carbonate ion? — add acid, will bubble (produce carbon dioxide)
18. What is the test for a sulfate ion? — add hydrochloric acid and barium chloride, white precipitate
19. What is a flame test? — when different metals are placed in a flame and produce colours.
20. What colour are the flames of potassium, calcium, and barium? — lilac, orange-red, green
21. Why can a flame test not be used to identify two metals mixed together? — the colours will mix

C6 Chemical analysis

Now use the questions below to check your knowledge from previous chapters.

C6

Previous questions | Answers

#	Previous questions	Answers
1	What is a compound?	a substance made from two or more different atoms bonded
2	Describe the structure of the atom.	dense nucleus of protons and neutrons with electrons orbiting around
3	Why do ionic compounds have high melting points?	strong electrostatic attraction between oppositely charged ions
4	When do ionic compounds conduct electricity?	when molten or in solution
5	How does the reactivity of the halogens change down the group?	becomes less reactive
6	Why does the reactivity of the halogens decrease down the group?	bigger atoms, outer shell further from nucleus, weaker electrostatic force of attraction, harder to gain an electron.
7	What is a displacement reaction?	when a more reactive element takes the place of a less reactive one in a compound
8	What name is given to the group 0 elements?	the noble gases
9	What do we call a reaction where an atom loses electrons?	oxidation
10	What are low-grade ores?	ores which are not financially worth extracting by normal means

(Put paper here)

Required Practical Skills

Practise answering questions on the required practicals using the example below.
You need to be able to apply your skills and knowledge to other practicals too.

Identifying ions	Worked example	Practice
You need to be able to describe how to identify unknown compounds, including: • flame tests for different ions • metal ion precipitation test • carbonate test • sulfate test • halide test. A common exam question asks you to identify mystery compounds based on their results in the above tests, so you need to know their methods and the observations for positive results. You also need to be able to write the formula for the ions these tests identify, and write equations using them.	A student wanted to test a sample for sulfate and halide ions. Write a method that the student could use to test for these two ions. **Answer**: Split the sample between two test tubes. In one test tube, carry out the test for sulfate ions by adding a few drops of dilute hydrochloric acid and dilute barium chloride solution. If a white precipitate appears, the sample contains sulfate ions. In the other test tube, carry out the test for halide ions by adding a few drops of dilute nitric acid followed by dilute silver nitrate solution. Formation of a white, yellow, or cream precipitate indicates the presence of halide ions.	1 When testing for halides it is a good idea to test three known samples to use as a reference. Explain why. 2 A sample produced a lilac colour in a flame test, and effervesced when treated with hydrochloric acid. Identify the formula of the sample. 3 Describe why the colour of the flame test for sodium is hard to see, and suggest a way to resolve this problem.

C6 Retrieval 65

Practice

Exam-style questions

01 A student pours dilute hydrochloric acid into a test tube and adds magnesium ribbon. A gas is formed that the student collects.

01.1 Name the **two** products formed in the reaction. **[2 marks]**

1 _____

2 _____

> **Exam Tip**
> The question has given you a clue to one of the products, make sure you've listed a gas.

01.2 Describe the test the student can carry out to identify the gas collected. Give the expected result. **[2 marks]**

test: _____

expected result: _____

01.3 Another student reacted magnesium with sodium carbonate. A gas was formed that the student collected.

Describe the test the student can carry out to identify the gas collected. Give the expected result. **[2 marks]**

test: _____

expected result: _____

01.4 The scientist had a sample of a pale green gas. They inserted a glowing splint into the gas. The splint went out.

The scientist then put some damp litmus paper into the gas. The litmus paper turned white.

Identify the gas. **[1 mark]**

_____

> **Exam Tip**
> Don't worry if you don't know which gas is pale green, it's the results of the test that are important.

02 A student has three metal compounds. They mix a sample of each compound with sodium hydroxide.

02.1 Metal compound **A** produces a white precipitate. When more sodium hydroxide solution is added, the precipitate dissolves.
Identify the metal. **[1 mark]**

_____

> **Exam Tip**
> You need to recall what happens when a small amount of sodium hydroxide is added, and what happens when it is in excess. There are two results needed for a reaction with sodium hydroxide.

02.2 Metal compound **B** produces a blue precipitate.
Identify the metal ion. **[1 mark]**

_____

02.3 Metal compound **C** produces a white precipitate. It does not dissolve on addition of further sodium hydroxide solution.
Identify the metal. **[1 mark]**

_____

C6

02.4 The student then carries out some tests to identify the non-metal ion in each compound. **Table 1** shows their results.

Table 1

Metal compound	Addition of barium chloride	Addition of dilute hydrochloric acid
A	white precipitate	no change
B	white precipitate	no change
C	white precipitate	no change

Identify the non-metal ion in all three compounds. **[1 mark]**

> **Exam Tip**
> It will be the same non-metal ion for all three.

02.5 Write a balanced symbol equation for the reaction in **02.2**. **[3 marks]**

03 All plants carry out photosynthesis. A student sets up the apparatus shown in **Figure 1** to investigate the products of photosynthesis.

Figure 1

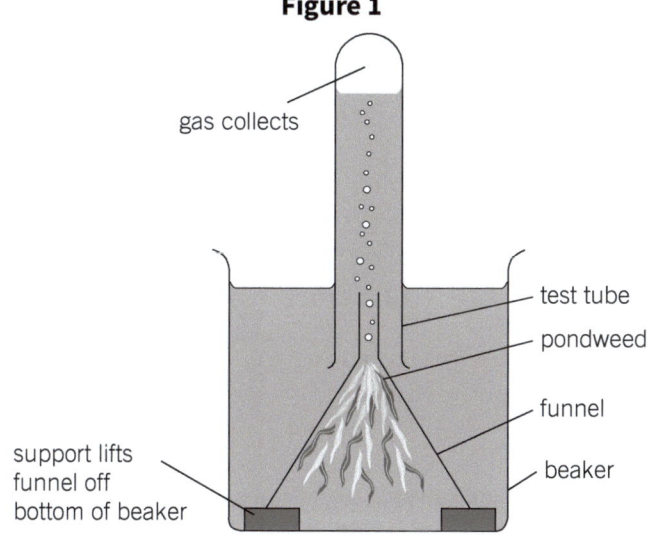

> **Exam Tip**
> Don't worry that photosynthesis is a topic in biology, the question isn't really asking you about that. The examiners want to see you apply your chemistry skills in a new context.

The student collects some of the gas in the test tube and inserts a glowing split. The splint relights.

03.1 Use the student's result to complete and balance the symbol equation for photosynthesis. **[2 marks]**

____ CO_2 + ____ H_2O → $C_6H_{12}O_6$ + ____

> **Exam Tip**
> Only fill in the gaps. If you're tempted to try to write an answer outside the gaps, STOP!

C6 Practice 67

03.2 Photosynthesis occurs in the leaves of plants. Pigments in the leaves help the process to occur. A student uses paper chromatography to investigate the pigments in a leaf. Describe a method to carry out the paper chromatography experiment. In your answer, name any equipment required. **[6 marks]**

03.3 The student finds that there are four pigments in the leaves. Sketch the chromatogram that the student has produced. **[1 mark]**

03.4 The student wants to calculate the R_f value of one of the pigments in the chromatogram. Give the equation to calculate the R_f value. **[1 mark]**

03.5 The solvent travelled 12.0 cm. One of the spots travelled 8.6 cm. Use **Table 2** to identify which pigment was responsible for the spot. **[2 marks]**

Table 2

Pigment	R_f value
carotene	0.95
xanthophyll	0.72
chlorophyll a	0.65
chlorophyll b	0.45

04 Chemists use different chemical tests to identify the substances in a compound.

04.1 A chemist had three unknown gases. The chemist carried out three simple tests to identify the gases. Their observations are shown in **Table 3**.

Table 3

Gas	Burning splint held at open end of tube	Glowing splint inserted into tube	Bubbled through limewater
A	no observation	no observation	cloudy
B	no observation	splint relights	no change
C	pop sound	no observation	no change

Identify the gases **A**, **B**, and **C**. **[3 marks]**

> **! Exam Tip**
> This is three marks for three short answers, don't be tempted to explain your reasoning because you won't get any extra marks and you'll just be wasting time.

04.2 The chemist has a fourth gas. The chemist thinks the gas is chlorine. Describe how the chemist could confirm that the gas is chlorine. **[2 marks]**

04.3 The chemist also has a sample of lithium bromide, magnesium bromide, and lithium carbonate. All three compounds are white solids. The compounds are not labelled.

Describe an experimental procedure the chemist could use to identify each compound. Your procedure should use as few tests as possible. Include the expected results for each test. **[6 marks]**

05 A student investigated the dyes in three felt tip pens. The dyes are soluble in water. They set up a chromatography experiment. **Figure 2** shows the apparatus the students used.

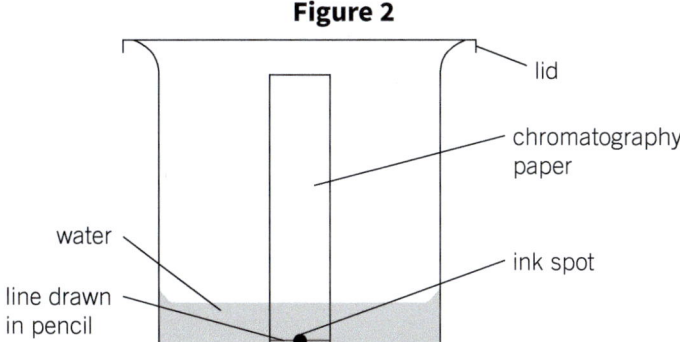

Figure 2

05.1 Name the mobile phase in this chromatography experiment.
[1 mark]

05.2 The student made a mistake in setting up the apparatus. Identify the mistake and give **one** problem caused by this mistake.
[2 marks]

> **Exam Tip**
> Look carefully at the diagram to find the mistake.

05.3 Another student set up the apparatus correctly. **Figure 3** shows the chromatogram the student obtained.

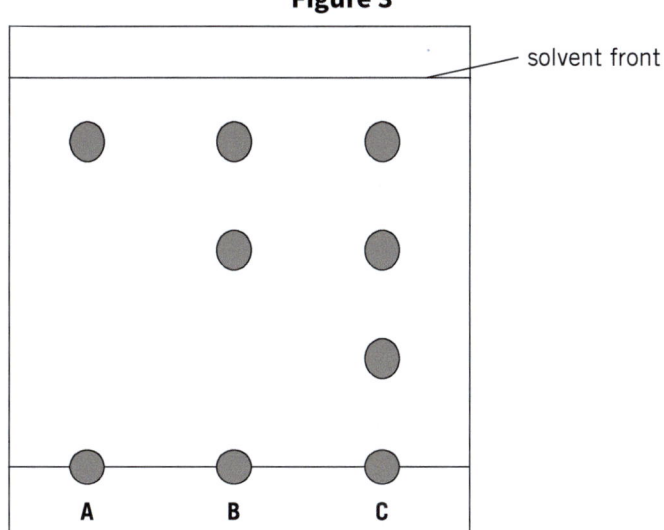

Figure 3

> **Exam Tip**
> Draw horizontal lines across from each spot so you can see which other ink samples they appear in.

Write **two** conclusions that compare the dyes in ink samples **A**, **B**, and **C**.
[2 marks]

05.4 Calculate the R_f value for the spot obtained from ink sample **A**.
[2 marks]

05.5 Circle the dye that is more attracted to the paper than all the other dyes. Justify your answer.
[2 marks]

06 A student had three unknown ionic compounds, **A**, **B**, and **C**.
The student carried out tests on the compounds to identify them.
Table 4 shows their observations.

Table 4

Compound	Tests with solution			Test with solid
	Add sodium hydroxide solution	Add silver nitrate solution	Add barium chloride solution	Add dilute hydrochloric acid
A	blue precipitate	no change	no change	bubbles observed
B	no change	white precipitate	no change	no change
C	brown precipitate	no change	precipitate, colour difficult to see	no change

06.1 Name the acid that should be added with silver nitrate. **[1 mark]**

06.2 Identify compound **A**. Justify your answer. **[3 marks]**

> **Exam Tip**
> Start with identifying the metal ion and then the non-metal ion.

06.3 Suggest a practical procedure the student could carry out to confirm the colour of the precipitate formed when barium chloride solution is added to compound **C**. **[1 mark]**

06.4 Write a conclusion about the identity of compound **B**. **[2 marks]**

06.5 Suggest a further test the student could carry out on compound **B** to help to identify the positive ion. **[1 mark]**

06.6 Use your answer to **06.5** to give **three** possible observations and a conclusion for each one. **[3 marks]**

07 A student has a mixture of coloured liquids. The student wants to identify which substances are in the mixture.

07.1 The student thinks that there are two substances dissolved in water. They carry out a chromatography experiment to identify whether this is accurate. They use water as the solvent. Sketch the chromatogram the student would see if there are two substances dissolved in water. **[3 marks]**

> **Exam Tip**
> Don't forget to label everything in your diagram.

07.2 The student finds that the mixture only contains one substance dissolved in water. They carry out another chromatography experiment to identify the substances. Their chromatogram is shown in **Figure 4**.

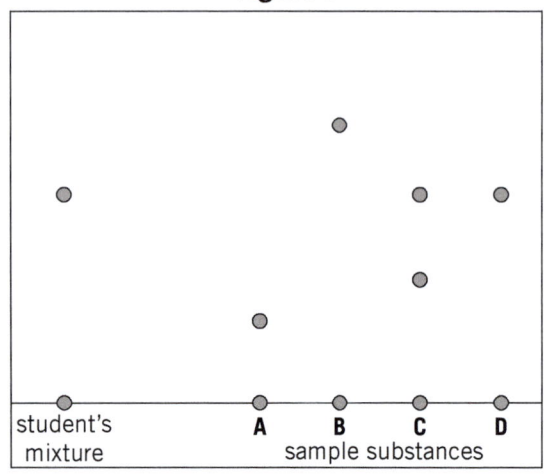

Figure 4

Identify the substance in the student's mixture. **[1 mark]**

07.3 The student wants to separate the substance from the water it is dissolved in. The substance has a boiling point of 78 °C. Explain why fractional distillation has to be used to extract the substance from the water. **[5 marks]**

08 For each of the following pairs of substances, suggest **one** chemical test that you can use to tell them apart. Give the result of the test for both substances.

08.1 sodium carbonate and sodium nitrate **[3 marks]**

08.2 potassium chloride and potassium iodide **[3 marks]**

08.3 calcium chloride and magnesium chloride **[3 marks]**

08.4 iron(II) sulfate and iron(III) sulfate **[3 marks]**

09 A student has a solution of an unknown aluminium compound. They only have a small sample of the solution, so they decide to carry out all three tests for the negative ion in the same test tube. The method they plan to use is:

1 Test for carbonate: add 1 cm³ of dilute sulfuric acid. Collect gas given off and pass through limewater.

2 Test for sulfates: add 1 cm³ of barium chloride and 1 cm³ hydrochloric acid.

3 Test for halides: add 1 cm³ of nitric acid and 1 cm³ of silver nitrate solution.

09.1 Explain why the suggested method will give a false result. **[2 marks]**

> **! Exam Tip**
> Think carefully about the order of the reagents.

09.2 Describe how the method can be improved to prevent false results. **[1 mark]**

09.3 Suggest **one** improvement to the method that would reduce the number of substances needed to carry out the tests. **[1 mark]**

09.4 Write a balanced symbol equation for the reaction that occurs in the test for carbonates. Assume the unknown compound is aluminium carbonate, $Al_2(CO_3)_3$. **[3 marks]**

10 A student has three metal compounds labelled **A**, **B**, and **C**. The student knows that they are a calcium compound, iron(II) sulfate, and an aluminium compound, but they do not know which is which.

10.1 Describe how the student can identify which metal compound is which. **[5 marks]**

10.2 The student identified compound **B** as the iron compound. Describe how the student can confirm that compound **B** is iron sulfate. **[3 marks]**

10.3 Write a balanced symbol equation for the reaction to identify the iron(II) in **10.1**. **[2 marks]**

10.4 The student identified that compound **A** was the calcium compound and that compound **C** was the aluminium compound. The student tested both compounds with silver nitrate solution. Compound **A** produced a yellow precipitate. Compound **C** produced a white precipitate. Identify the **two** compounds. **[2 marks]**

11 A scientist carried out a series of tests to identify an unknown metal compound.

11.1 First the scientist wanted to identify the non-metal ion. They carried out three chemical tests. **Table 5** shows their observations.

Table 5

Chemical test	Observation
add barium chloride solution and hydrochloric acid	white precipitate formed
add silver nitrate solution and nitric acid	no precipitate
add dilute acid then collect and test gas	solution remained clear and colourless

Describe how the scientist tested the gas in the third chemical test. **[1 mark]**

11.2 Identify the non-metal ion in the metal compound. **[1 mark]**

11.3 To identify the metal, the scientist carried out a flame test on the metal compound. The scientist was unable to identify one colour in the flame test. Suggest why. **[1 mark]**

12 A student wanted to look at the different compounds in a mixture. The student set up the paper chromatography experiment in **Figure 5** to separate out the compounds.

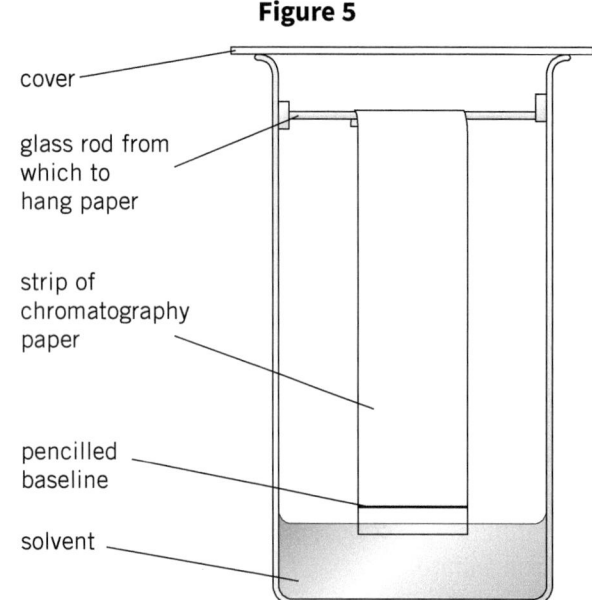

Figure 5

12.1 Describe why the student drew the baseline in pencil and not in pen. **[1 mark]**

12.2 Describe why the student made sure that the solvent was below the baseline. **[1 mark]**

12.3 Describe why the student placed a lid on top of the beaker. **[1 mark]**

> **Exam Tip**
> Sketch means you don't need to include values but you do still need to use a ruler and label everything.

12.4 The mixture was made of three compounds. Compound **A** is a pure substance. Compound **B** is a mixture of **A** and **C**. Compound **C** is a pure substance that is different to **A**. Sketch the chromatogram that the student produced. **[3 marks]**

12.5 The R_f value can be used to determine the identity of a substance in chromatography. Give the equation to calculate the R_f value of a substance. **[1 mark]**

12.6 The student calculated the R_f value of substance **C** to be 0.31. Use **Table 6** to identify substance **C**. **[1 mark]**

Table 6

Substance	R_f value
methyl red	0.30
ethyl green	0.46
titan yellow	0.61
fuchsin acid	0.89

> **Exam Tip**
> None of the values in the table exactly match the value calculated by the student, but often an experimentally-obtained value won't be identical to the value from a database.

13 Sodium reacts with chlorine to form sodium chloride.

13.1 Explain how the structure and bonding in sodium and chlorine give rise to their different properties. **[5 marks]**

13.2 Sodium chloride is an ionic compound. Describe the structure and bonding in sodium chloride. Explain why this compound has a high melting point. **[6 marks]**

> **Exam Tip**
> Before you start, determine the charge on the ions of sodium and chlorine.

13.3 Write a balanced symbol equation for the reaction between sodium and chlorine. Include state symbols. **[3 marks]**

14 A chemist tried to pass an electric current through a solid, a liquid, and a solution. **Table 7** shows the chemist's results.

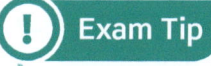

Table 7

Substance	State	Observations at anode	Observations at cathode
sodium chloride	solid	no change (did not conduct electricity)	
sodium chloride	liquid	smell of chlorine	silver-coloured liquid produced
sodium chloride	concentrated solution	gas produced did not relight glowing splint smell of chlorine	gas produced lit splint gives a squeaky pop
sodium chloride	dilute solution	gas produced relit glowing splint smell of chlorine	gas produced lit splint gives a squeaky pop

14.1 Explain the observations in solid and liquid sodium chloride. **[3 marks]**

14.2 Write a half equation, including state symbols, for the reaction that occurs at the cathode for concentrated sodium chloride. **[3 marks]**

> **Exam Tip**
> There are two gases produced here, not one gas that gives two positive results.

14.3 Suggest an explanation for the observations at the anode and cathode for dilute sodium chloride solution. **[6 marks]**

Knowledge

C7 Acids, bases, and salts

Acids and alkalis

Acids are compounds that, when dissolved in water, release H^+ ions. There are three main acids: sulfuric acid, H_2SO_4, nitric acid, HNO_3, and hydrochloric acid, HCl.

Alkalis are compounds that, when dissolved in water, release OH^- ions. The **pH** scale is a measure of acidity and alkalinity. It runs from 1 to 14.
- Aqueous solutions with pH < 7 are acidic.
- Aqueous solutions with pH > 7 are alkaline.
- Aqueous solutions with pH = 7 are **neutral**.

Indicators

Indicators can show if something is an acid or an alkali.
- **Universal indicator** can also tell us the approximate pH of a solution.
- Electronic pH probes can give us the exact pH of a solution.

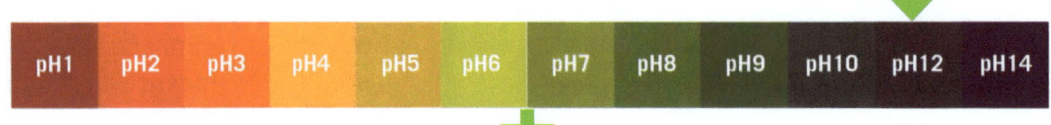

Salts

When acids react with metals or metal compounds, they form **salts**. A salt is a compound where the hydrogen from an acid has been replaced by a metal. For example, nitric acid, HNO_3, reacts with sodium to form $NaNO_3$. The H in nitric acid is replaced with Na.

The table shows how to name salts.

Acid	hydrochloric acid	sulfuric acid	nitric acid
Formula	HCl	H_2SO_4	HNO_3
Ions formed in solution	H^+ and Cl^-	$2H^+$ and SO_4^{2-}	H^+ and NO_3^-
Type of salt formed	metal chloride	metal sulfate	metal nitrate
Sodium salt example	sodium chloride, NaCl	sodium sulfate, Na_2SO_4	sodium nitrate, $NaNO_3$

Making soluble salts

Soluble salts are salts that dissolve. Soluble salts can be made by reacting an acid with one of these different substances:

- Metals – choose the metal carefully because some metals are too reactive, e.g., potassium, and others are not reactive enough, e.g., gold.
- Insoluble bases – add the base to the acid until no more will react. Then filter the solution to remove the excess solid.
- Alkalis – add the alkali to the acid gradually and use an indicator to show when the acid and alkali have completely reacted. For example, here is a method for making a soluble salt using an acid, alkali and indicator:

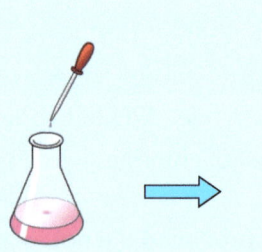  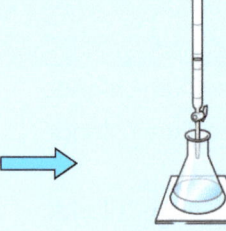

1 Put the alkali in the flask with a few drops of indicator.

2 Add acid from the burette until the indicator changes colour. You record the volume of acid added.

3 Repeat the experiment without indicator in the flask. Add the volume of acid you recorded in the last step.

5 Allow crystals to form. You then filter these. The crystals can then be carefully washed and dried with filter paper.

4 Put the solution from the flask into an evaporating basin and evaporate the water until the crystallisation point is reached.

C7

Reactions of acids

Reactions of acids with metals
Acids react with some metals to form salts and hydrogen gas.

magnesium + hydrochloric acid → sodium chloride + hydrogen

Reactions of acids with metal hydroxides
Acids react with metal hydroxides to form salts and water.

hydrochloric acid + sodium hydroxide → sodium chloride + water

The ionic equation for this reaction is always:

$$H^+(aq) + OH^-(aq) \rightarrow H_2O(l)$$

Reactions of acids with metal oxides
Acids react with metal oxides to form salts and water.

hydrochloric acid + sodium oxide → sodium chloride + water

Reactions of acids with metal carbonates
Acids react with metal carbonates to form a salt, water, and carbon dioxide.

hydrochloric acid + sodium carbonate → sodium chloride + water + carbon dioxide

neutralisation reactions

Redox

The reaction of an acid with a metal is a redox reaction:
- The metal loses electrons – it is oxidised.
- Hydrogen gains an electron – it is reduced.

Alkalis and bases

Bases neutralise acids to form water in **neutralisation** reactions. Some metal hydroxides dissolve in water to form alkaline solutions, called alkalis. Another alkali is ammonia, which dissolves in water and will react with an acid to form an ammonium salt.

Some metal oxides and metal hydroxides do not dissolve in water. They are bases, but are not alkalis. Calcium hydroxide is slightly soluble in water. A solution of it is called limewater and it readily reacts with carbon dioxide to produce calcium carbonate.

Crystallisation

You can produce a solid salt from an insoluble base by **crystallisation**.

The experimental method is:
1. Choose the correct acid and base to produce the salt.
2. Put some of the dilute acid into a flask. Heat gently with a Bunsen burner.
3. Add a small amount of the base and stir.
4. Keep adding the base until no more reacts – the base is now in excess.
5. Filter to remove the unreacted base.
6. Add the remaining solution to an evaporating dish.
7. Use a water bath or electric heater to evaporate the water. The salt crystals will be left behind.

Making insoluble salts

Some salts do not need to be crystallised because they are naturally insoluble.

When the salt forms it becomes a **precipitate**, which can be filtered off.

This is a good way of removing ions from water, for example, in treating drinking water or **effluent**.

Key Terms
Make sure you can write a definition for these key terms.

| acid | alkali | base | crystallisation | effluent | indicator | ionise | neutral |
| neutral | neutralisation | pH | salt | precipitation | universal indicator |

C7 Knowledge

Retrieval

Learn the answers to the questions below then cover the answers column with a piece of paper and write as many as you can. Check and repeat.

C7 questions | Answers

#	Question	Answer
1	In terms of pH, what is an acid?	a solution with a pH of less than 7
2	In terms of pH, what is a neutral solution?	a solution with a pH of 7
3	In terms of H^+ ions, what is an acid?	a substance that releases H^+ ions when dissolved in water
4	How is the amount of H^+ ions in a solution related to its pH?	the more H^+ ions, the lower the pH
5	Give the names and formulae of three main acids.	hydrochloric acid, HCl; sulfuric acid, H_2SO_4; nitric acid, HNO_3
6	How can we tell if something is an acid?	by using an indicator or pH probe
7	What is the ionic equation for a reaction between an acid and an alkali?	$H^+(aq) + OH^-(aq) \rightarrow H_2O(l)$
8	How can you obtain a solid salt from a solution?	crystallisation
9	What is a salt?	compound formed from a reaction of a metal or metal containing compound with an acid
10	Which type of salts do sulfuric acid, hydrochloric acid, and nitric acid form?	sulfates, chlorides, nitrates
11	What are the products of a reaction between a metal and an acid?	salt + hydrogen
12	What are the products of a reaction between a metal hydroxide and an acid?	salt + water
13	What are the products of a reaction between a metal oxide and an acid?	salt + water
14	What are the products of a reaction between a metal carbonate and an acid?	salt + water + carbon dioxide
15	What is a base?	substance that reacts with acids in neutralisation reactions
16	What is an alkali?	substance that dissolves in water to form a solution above pH 7
17	What is a neutralisation reaction?	a reaction between an acid and a base to produce water
18	Name two uses for precipitation and filtration.	treating drinking water and effluent

C7 Acids, bases, and salts

C7

Now use the questions below to check your knowledge from previous chapters.

Previous questions | Answers

#	Question	Answer
1	Describe the structure and bonding of small molecules.	small groups of atoms group together into molecules with strong covalent bonds between the atoms, weak intermolecular force between the molecules
2	Describe the structure and bonding of large molecules such as polymers.	many identical molecules all joined together by strong covalent bonds in a long chain, with weak intermolecular forces between the chains
3	Explain why giant covalent substances have high melting points.	strong covalent bonds between atoms require a lot of energy to break
4	What is electrolysis?	the process of using electricity to extract elements from a compound
5	What is the cathode?	the negative electrode
6	What name is given to the Group 0 elements?	the noble gases
7	Why are the noble gases inert?	they have full outer shells, do not need to lose or gain electrons
8	How does the melting point of the noble gases change down the group?	increases
9	What is an atom?	the smallest part of an element that can exist
10	What is an element?	a substance made of one type of atom

Maths Skills

Practise your maths skills using the worked example and practice questions below.

Lines of best fit

When describing lines of best fit, you need to state

- its correlation
- if the line is straight or curved
- whether the line plateaus (stops changing and flattens out)
- whether the line runs through the origin (0,0).

Correlations can be positive or negative and either strong or weak, or there can be no correlation.

If the line of best fit is straight and goes through the origin, the variables are **directly proportional** to each other, meaning as one variable changes the other changes at the same rate.

Worked example

Fully describe the curved line of best fit for the reaction plotted on the graph below.

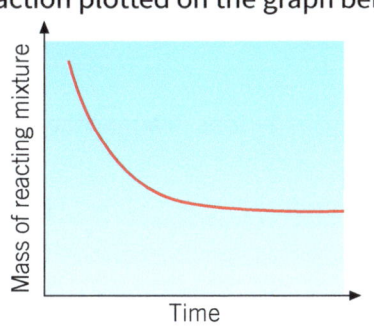

Answer:

The graph shows a negative correlation.

The curved line of best fit does not pass through the origin.

The mass of the mixture decreases rapidly at first, this decrease then slows down in the middle of the reaction, and finally plateaus as the mass stops decreasing with time.

Practice

The graph below shows how the volume of gas produced changes with time in the reaction between marble chips and hydrochloric acid.

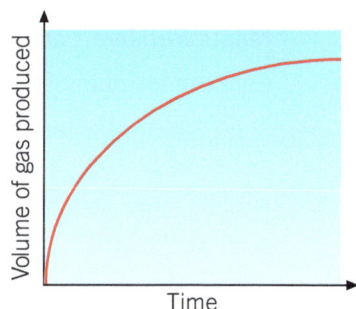

1 Describe the graph.

2 Sketch a graph to show a directly proportional relationship between two variables.

Practice

Exam-style questions

01 A student has two solutions of hydrochloric acid.
Table 1 shows the pH of each solution.

Table 1

Solution	pH
A	1
B	2

01.1 Identify which acid, **A** or **B**, is more acidic.
Give a reason for your answer. **[1 mark]**

_____

01.2 Name the products of the reaction of copper carbonate with hydrochloric acid. **[1 mark]**

_____

> **! Exam Tip**
> There will be **three** products from this reaction – it is important that you can recall and apply all of the general salt equations in chemistry.

01.3 Give the chemical formula of the salt produced in the reaction between hydrochloric acid and copper carbonate. **[1 mark]**

_____

02 Sulfuric acid reacts with magnesium to form magnesium sulfate.

02.1 The chemical formula of sulfuric acid is H_2SO_4.
Give the charge of the sulfate ion, SO_4, in sulfuric acid. **[1 mark]**

_____

02.2 Give the chemical formula of magnesium sulfate. **[1 mark]**

> **! Exam Tip**
> Sulfuric acid is a neutral compound so the charges on the ions within sulfuric acid must be equal to zero.

_____

02.3 Manganese is another metal that forms 2+ ions. Manganese reacts with hydrochloric acid.
Give the formula of manganese chloride. **[1 mark]**

_____

02.4 Name the gas released when magnesium and manganese react with acids. **[1 mark]**

_____

03 This question is about the pH scale. A student measured the pH of some solutions. **Table 2** shows the results the student obtained.

Table 2

Solution	pH
A	7
B	2
C	10
D	5
E	12

03.1 Name **two** ways of measuring the pH of a solution. **[2 marks]**

03.2 Give the letter of the neutral solution. **[1 mark]**

03.3 Give the letter of the most alkaline solution. **[1 mark]**

03.4 Give the letter of the solution that has the highest concentration of hydrogen ions, H^+. **[1 mark]**

03.5 Some alkali is added to solution **A**. Write down whether the pH increases, decreases, or stays the same. **[1 mark]**

04 A student wanted to make sodium chloride crystals from sodium hydroxide and hydrochloric acid solutions:

$$NaOH(aq) + HCl(aq) \rightarrow NaCl(aq) + H_2O(l)$$

The student used the following method:
1. Use a measuring cylinder to transfer 25 cm³ of sodium hydroxide into a conical flask.
2. Add 1 cm³ of indicator.
3. Add dilute hydrochloric acid from a burette until the indicator changes colour.
4. Pour the mixture from the flask into a beaker.
5. Heat the beaker and its contents until half the water has evaporated.
6. Allow the rest of the water to evaporate by leaving the beaker in a warm, dry place.

> **Exam Tip**
>
> Can you think of a better way to measure these?

04.1 Suggest and explain **one** improvement to step **1** and **one** improvement to step **2**. **[4 marks]**

04.2 Describe what the students should do between steps **3** and **4**. Give a reason for this extra step. **[2 marks]**

05 Table 3 shows the strengths of the covalent bonds in six molecules.

Table 3

Element	Formula of molecule	Bond strength in kJ/mol
nitrogen	N_2	944
oxygen	O_2	496
hydrogen	H_2	436
chlorine	Cl_2	242
bromine	Br_2	193
iodine	I_2	151

05.1 Suggest a reason for the difference in bond strengths for Cl_2, Br_2, and I_2. Use the Periodic Table to help you answer this question.
[2 marks]

05.2 Explain the difference in bond strengths for N_2, O_2, and H_2.
Include dot and cross diagrams in your answer. **[4 marks]**

06 The formula of ammonia is NH_3. **Figure 1** is a ball and stick model of an ammonia molecule.

Figure 1

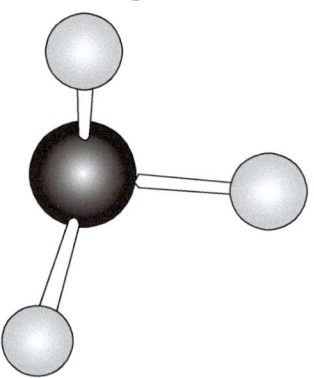

06.1 The ball and stick model is not a true representation of ammonia.
Give **one** reason why. **[1 mark]**

06.2 Draw a dot and cross diagram to show the covalent bonds in ammonia. **[2 marks]**

> **! Exam Tip**
>
> Covalent bonding has overlapping circles – ionic bonding is the one with square brackets.

06.3 Explain why ammonia does not conduct electricity. **[1 mark]**

06.4 **Table 4** shows the boiling points and formulae of two compounds.

Table 4

Name of compound	Formula	Boiling point in °C
ammonia	NH_3	−334
hydrazine	N_2H_4	114

Give the state of ammonia at 20 °C. **[1 mark]**

06.5 Explain why hydrazine has a higher boiling point than ammonia. **[2 marks]**

> **! Exam Tip**
>
> You've probably never heard of hydrazine before, but this is just a new context to apply your knowledge to.

C7 Acids, bases, and salts

07 This question is about the reactivity series of metals.

07.1 Compare the reactions of potassium, lithium, and magnesium with cold water. **[6 marks]**

07.2 Caesium is at the bottom of Group 1 of the Periodic Table. Predict the names of the products of the reaction of caesium with water. **[1 mark]**

> **Exam Tip**
> Base your answer on what you know about sodium or potassium.

07.3 Explain how the reactivities of the Group 1 elements are related to the tendency of the elements to form their positive ions. **[2 marks]**

08 A teacher makes sodium chloride by adding burning sodium to a container of chlorine gas.

08.1 Suggest **one** safety precaution the teacher should take. **[1 mark]**

08.2 Balance the symbol equation for the reaction and add state symbols. **[2 marks]**

___ Na ___ + Cl$_2$ ___ → ___ NaCl ___

08.3 Describe the structure and bonding in solid sodium chloride. In your answer, outline how the ions are made and give their charges. **[6 marks]**

> **Exam Tip**
> To begin it might help to think about:
> - how many electrons sodium has in its outer shell and what happens to them
> - the charge on the sodium ions
> - how many electrons chlorine has in its outer shell and what happens to them
> - the charge on the chloride ions.

09 This question is about the formation of salts.

09.1 Complete the following word equations to show the formations of different salts.

sulfuric acid + magnesium hydroxide → ___ + ___

calcium + ___ → calcium chloride + ___

___ + nitric acid → potassium ___ + hydrogen

09.2 Calcium carbonate, CaCO$_3$, reacts with sulfuric acid, H$_2$SO$_4$, to produce calcium sulfate, CaSO$_4$, and two other products. Complete the balanced symbol equation for this reaction. **[2 marks]**

CaCO$_3$ + ___ → CaSO$_4$ + ___ + ___

> **Exam Tip**
> Look back at the general equations for the formation of different salts to help you work out the product and reactants.

09.3 Name the salt produced when sodium hydroxide reacts with sulfuric acid. **[1 mark]**

> **Exam Tip**
> Start by adding in the formula for sulfuric acid, then work out the rest of the products from the relevant salt equation.

09.4 A student added sodium hydroxide to sulfuric acid 1 cm³ at a time. After each addition, they recorded the pH of the solution. **Figure 2** shows their results.

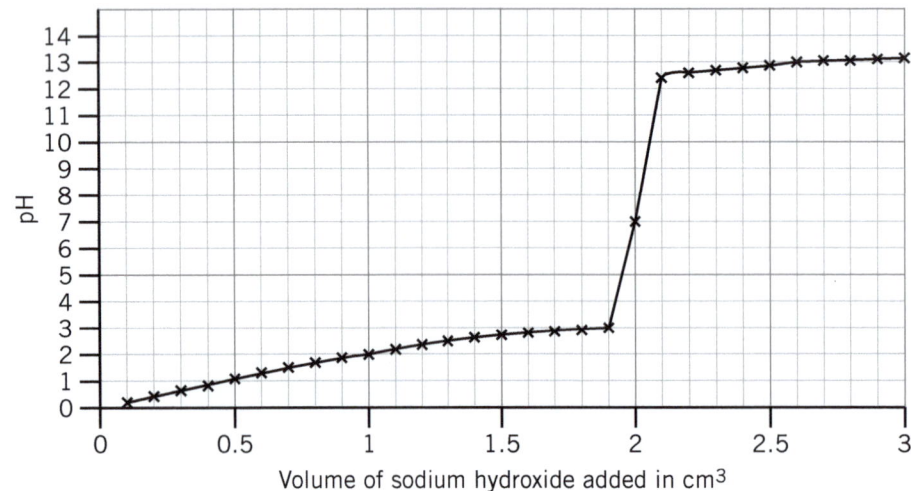

Determine the volume of sodium hydroxide that produced a neutral pH. **[1 mark]**

> **! Exam Tip**
> Draw lines on the graph to show your working, this will help prevent mistakes and show the examiner you know what you're doing!

10 Crude oil is a vital substance that is a mixture of hydrocarbons.

10.1 Explain what a hydrocarbon is. **[1 mark]**

10.2 A hydrocarbon was found to consist of 14.3% hydrogen and 85.7% carbon. Calculate the empirical formula of the hydrocarbon.
A_r: H = 1; C = 12 **[3 marks]**

10.3 Hydrocarbons are often used as fuels. Balance the symbol equation below that shows the complete combustion of Ethane.

___ C_2H_4 + ___ O_2 → ___ CO_2 + ___ H_2O **[2 marks]**

10.4 Hydrocarbons can also undergo incomplete combustion. Define incomplete combustion. **[1 mark]**

10.5 Some hydrocarbons contain sulfur impurities. Name the gas produced from these impurities and name the environmental problem this causes. **[2 marks]**

10.6 The length of the carbon chain effects the physical properties of the hydrocarbon. Complete **Table 5** to describe how increasing the length of the carbon chain affects each named property. **[3 marks]**

Table 5

Physical property	Effect increasing the carbon chain has on the property
boiling point	
flammability	
viscosity	

10.7 Name the process that is used to separate the different fractions of crude oil. **[1 mark]**

82   C7 Acids, bases, and salts

11 Sulfuric acid reacts with calcium to produce calcium sulfate and hydrogen gas.

11.1 Describe the test that would need to be carried out to prove the gas was hydrogen. Give the observation of a positive result. **[2 marks]**

11.2 The formula for sulfuric acid is H_2SO_4. Calculate the percentage mass of sulfur in sulfuric acid. A_r: H = 1; S = 32; O = 16 **[2 marks]**

11.3 The symbol equation for the reaction is
$$H_2SO_4 + Ca \rightarrow CaSO_4 + H_2$$
8.0g of Ca was completely reacted with excess sulfuric acid. The hydrogen gas was collected. Calculate the volume of gas, in dm^3, the hydrogen took up at room temperature and pressure.
A_r: Ca = 40; H = 1; S = 32; O = 16 **[2 marks]**

11.4 A student made up a solution of sulfuric acid by dissolving 0.200 moles of sulfuric acid in 250 cm^3 of distilled water. Calculate the concentration in mol/dm^3. **[2 marks]**

11.5 Describe how a student could prove the identity of the metal in calcium sulfate using a flame test. Include the positive result. **[3 marks]**

12 This question is about metals and the Periodic Table.

12.1 Below is a list of physical properties of Group 1 and transition metals. Choose all the answers that describe the properties of transition metals. **[2 marks]**
- low melting point
- high density
- form coloured compounds
- react vigorously with water
- soft

12.2 Sodium and lithium are both Group 1 elements. They both react with water to produce a hydroxide and hydrogen. Explain why sodium is more reactive than lithium. **[6 marks]**

12.3 Balance the equation to show the reaction of potassium and water.
$$\_\_ K + \_\_ H_2O \rightarrow \_\_ K_2O + \_\_ H_2$$ **[1 mark]**

12.4 Explain why potassium must be stored under oil. **[1 mark]**

12.5 The modern Periodic Table arranges the elements in order of atomic number. Name the scientist whose work led to the formation of the modern Periodic Table. **[1 mark]**

Knowledge

C8 Quantitative chemistry A

Conservation of mass

The **conservation** of mass states that atoms cannot be created or destroyed in a chemical reaction. Atoms are rearranged into new substances. All the atoms you had in the reactants must be present in the products.

As such, when it comes to measuring the mass of a reaction, you would expect the mass at the start to be the same as the mass at the end. However, sometimes the mass can appear to change.

Decrease in mass

In some reactions the mass appears to decrease. This is normally because a gas is produced in the reaction and lost to the surroundings. For example:

sodium + water → sodium hydroxide + hydrogen
$$2Na(s) + 2H_2O(l) \rightarrow 2NaOH(aq) + H_2(g)$$

The mass of the sodium and the water at the start of the reaction will be more than the mass of the sodium hydroxide at the end of the reaction, because hydrogen atoms have been lost as a gas.

Increase in mass

In some reactions the mass appears to increase. This is normally because one of the reactants is a gas. For example:

sodium + chlorine → sodium chloride
$$2Na(s) + Cl_2(g) \rightarrow 2NaCl(s)$$

The mass of the sodium at the start of the reaction will be lower than the mass of sodium chloride at the end of the reaction. This is because atoms from the gaseous chlorine have been added to the sodium, increasing the mass.

Balancing symbol equations

When writing symbol equations you need to ensure that the number of each atom on each side is equal.

$$2H_2 + O_2 \rightarrow 2H_2O$$

balanced

there are 4 hydrogen atoms on each side, and 2 oxygen atoms on each side

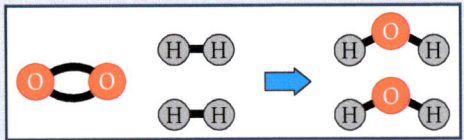

State symbols

A balanced symbol equation should also include **state** symbols.

State	Symbol
solid	(s)
liquid	(l)
gas	(g)
aqueous or dissolved in water	(aq)

Formula mass

Every substance has a **formula mass**, M_r.

M_r = sum (relative atomic mass of all the atoms in the formula)

Avogadro's constant

One mole of a substance contains 6.02×10^{23} atoms, ions, or molecules. This is **Avogadro's constant**.

One mole of a substance has the same mass as the M_r of the substance. For example, the M_r (H_2O) = 18, so 18 g of water molecules contains 6.02×10^{23} molecules, and is called one mole of water.

You can write this as: $\text{moles} = \dfrac{\text{mass}}{M_r}$

Ratios

In the reaction between hydrogen and oxygen, the ratio of hydrogen to oxygen molecules is 2:1. This means that for every *one* molecule of oxygen, you would need *two* molecules of hydrogen, for example:

- If you had 10 molecules of oxygen you would need 20 molecules of hydrogen.
- If you had one mole of oxygen you would need two moles of hydrogen.

A balanced symbol equation shows the ratios of the reactants and products in a chemical reaction.

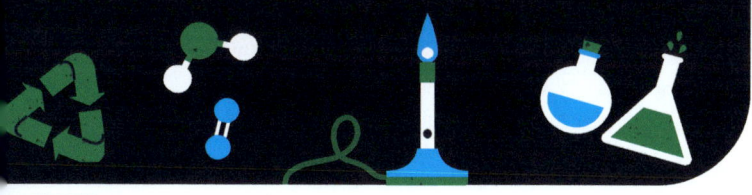

C8

Using balanced equations

In a balanced symbol equation the sum of the M_r of the reactants equals the sum of the M_r of the products.

If you are asked what mass of a product will be formed from a given mass of a specific reactant, you can use the steps below to calculate the result:
1. Balance the symbol equation.
2. Calculate moles of the substance with a known mass using $\text{moles} = \dfrac{\text{mass}}{M_r}$
3. Using the balanced symbol equation, work out the number of moles of the unknown substance.
4. Calculate the mass of the unknown substance using $\text{mass} = \text{moles} \times M_r$

If you are asked to balance an equation, you can use the steps below to work out the answer:
1. Work out M_r of all the substances.
2. Calculate the number of moles of each substance in the reaction using $\text{moles} = \dfrac{\text{mass}}{M_r}$
3. Convert to a whole number ratio.
4. Balance the symbol equation.

Percentage mass

Calculate the percentage mass of an element in a compound using the following equation:

% mass = relative atomic mass of element × $\dfrac{\text{number of atoms in molecule}}{\text{relative formula mass of compound}}$

Theoretical yield

The **theoretical yield** of a chemical reaction is the mass of a product that you expect to be produced.

Even though no atoms are gained or lost during a chemical reaction, it is not always possible to obtain the theoretical yield because:
- some of the product can be lost when it is separated from the reaction mixture
- there can be unexpected side reactions between reactants that produce different products
- the reaction may be reversible.

Empirical formula

The empirical formula of a compound is the simplest whole-number ratio of its elements.

For each element in the compound:

$\text{number of moles}, n = \dfrac{\text{mass of element}}{\text{mass of 1 mole of element}}$

This tells you how many moles of each element are present in the compound.

Use this to work out the simplest ratio of different elements in the compound.

This then gives you the empirical formula.

For example, if a hydrocarbon contains 75 g of carbon and 25 g of hydrogen:

moles of carbon = $\dfrac{75}{12}$ = 6.25

moles of hydrogen = $\dfrac{25}{1}$ = 25

So the ratio of carbon atoms to hydrogen atoms is 6.25:25. The simplest whole number ratio is 1:4, so the empirical formula is CH_4.

C8 Knowledge

Knowledge

C8 Quantitative chemistry B

Concentration

Concentration is the amount of solute in a volume of solvent.

The unit of concentration is g/dm³. Concentration can be calculated using:

$$\text{concentration (g/dm}^3\text{)} = \frac{\text{mass (g)}}{\text{volume (dm}^3\text{)}}$$

Sometimes volume is measured in cm³:

$$\text{volume (dm}^3\text{)} = \frac{\text{volume (cm}^3\text{)}}{1000}$$

- lots of solute in little solution = high concentration
- little solute in lots of solution = low concentration

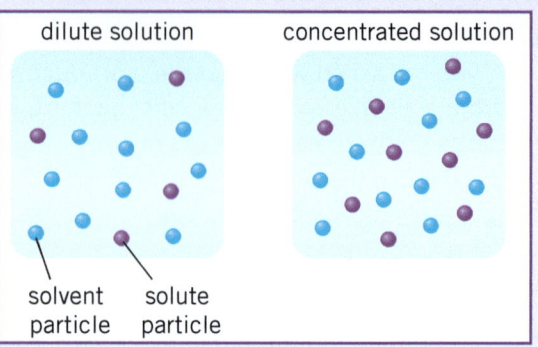

Concentration in mol/dm³

Concentration can also be measured in mol/dm³:

$$\text{concentration of solution (mol/dm}^3\text{)} = \frac{\text{number of moles of solute}}{\text{volume of solution (dm}^3\text{)}}$$

You can use this formula and $\text{mass} = \text{moles} \times M_r$ to calculate the mass of solute dissolved in a solution.

- The greater the mass of solute in solution, the greater the number of moles of solute, and therefore the greater the concentration.
- If the same number of moles of solute is dissolved in a smaller volume of solution, the concentration will be greater.

Moles of gases

At any given temperature and pressure, the same number of moles of a gas will occupy the same volume.

At room temperature (25 °C) and pressure (1 atm), one mole of *any* gas will occupy 24 dm³.

Calculating moles of gases

To calculate the number of moles of a gas:

$$\text{moles of a gas} = \frac{\text{volume (dm}^3\text{)}}{24\,\text{dm}^3}$$

or

$$\text{moles of a gas} = \frac{\text{volume (cm}^3\text{)}}{24\,000\,\text{cm}^3}$$

For example, to calculate the number of moles of 3 dm³ of carbon dioxide at room temperature and pressure:

$$\text{number of moles} = \frac{3}{24} = 0.125\,\text{mol}$$

mol is the unit of moles

Volume of gases

0.125 moles of carbon dioxide, oxygen, and hydrogen will all have the volume 3 dm³.

To calculate the volume of a gas at room temperature and pressure:

$$\text{volume (dm}^3\text{)} = \text{moles} \times 24\,\text{dm}^3$$

For example, to calculate the volume of 0.25 mol of chlorine at room temperature and pressure:

$$\text{volume (dm}^3\text{)} = 0.25\,\text{mol} \times 24\,\text{dm}^3 = 6\,\text{dm}^3$$

Indicators

We use an indicator so we know when the right amount of acid or alkali has been added during a titration. When the indicator changes colour we stop adding acid or alkali from the burette.

Indicator	Colour in acid	End point	Colour in alkali
methyl orange	red	orange	yellow
phenolphthalein	colourless	pale pink	pink

C8

Titration

Titration is an experimental technique to work out the concentration of an unknown solution in the reaction between an acid and an alkali.

1. Use a **pipette** to extract a known volume of the solution with an unknown concentration. A pipette measures a fixed volume only.
2. Add the solution of unknown concentration to a conical flask and put the conical flask on a white tile.
3. Add a few drops of a suitable indicator to the conical flask.
4. Add the other solution with a known concentration to the **burette**.
5. Carry out a rough titration to find out approximately what volume of solution in the burette needs to be added to the solution in the conical flask. Add the solution from the burette to the solution in the conical flask $1\,cm^3$ at a time until the end point is reached.
6. The **end point** is when the indicator has just changed colour.
7. Record the volume of the end point as your rough value.
8. Now repeat steps **1–7**, but as you approach the end point add the solution from the burette drop-by-drop. Swirl the conical flask in between drops.
9. Record the volume of the end point.

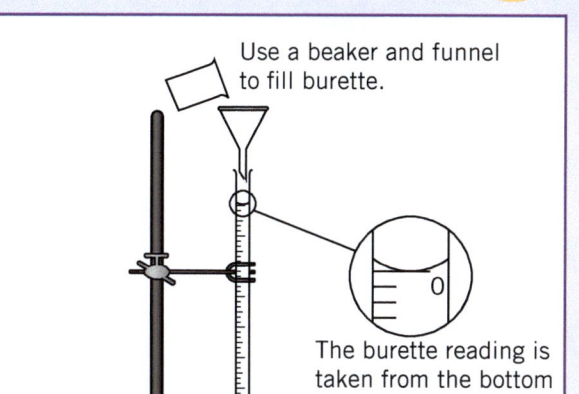

Use a beaker and funnel to fill burette.

The burette reading is taken from the bottom of the meniscus.

Use one hand to control the flow rate.

Swirl the flask with the other hand whilst the drops are being added.

Concordance

Repeat the titration until you get **concordant titres**.

- A titre is the volume of solution that you have added from your burette.
- Concordant means that the titres are within $0.1\,cm^3$ of each other.

You would stop when you had two concordant results, and calculate the mean:

$$\text{mean} = \frac{\text{sum of the concordant results}}{\text{number of concordant results}}$$

Calculating concentration

To calculate the concentration of the unknown solution (the solution in the conical flask):

1. Write a balanced symbol equation for the reaction.
2. Calculate the moles used from the known solution using:

   $\text{moles} = \text{concentration (mol/dm}^3\text{)} \times \text{volume (dm}^3\text{)}$

3. Use the ratio from the balanced symbol equation to deduce the number of moles present in the unknown solution.
4. Calculate the concentration of the unknown solution using: $\text{concentration (mol/dm}^3\text{)} = \dfrac{\text{moles}}{\text{volume (dm}^3\text{)}}$

Key Terms

Make sure you can write a definition for these key terms.

Avogadro's constant	balanced	burette	calculation	concentration
concordant	conservation	end point	formula mass	mole
pipette	state	theoretical yield	titration	titre

Retrieval

Learn the answers to the questions below then cover the answers column with a piece of paper and write as many as you can. Check and repeat.

C8 questions | Answers

#	Question	Answer
1	What is the conservation of mass?	that atoms cannot be created or destroyed
2	When a metal forms a metal oxide, why does the mass increase?	because oxygen atoms have been added
3	When an acid reacts with a metal, why does the mass decrease?	because a gas is produced and escapes
4	What is relative formula mass?	the sum of the relative atomic masses of each atom in a compound
5	What are the four state symbols and what do they stand for?	(s) solid, (l) liquid, (g) gas, (aq) aqueous
6	What symbol do we use for relative formula mass?	M_r
7	What is a mole?	a number of particles
8	Give the value for Avogadro's constant.	6.022×10^{23}
9	What formula relates moles, mass, and M_r?	$\text{moles} = \dfrac{\text{mass}}{M_r}$
10	What is a unit for concentration?	g/dm^3 or mol/dm^3
11	Which formula relates concentration, mass, and volume?	$\text{concentration (g/dm}^3\text{)} = \dfrac{\text{mass (g)}}{\text{volume (dm}^3\text{)}}$
12	If the amount of solute in a solution is increased, what happens to its concentration?	increases
13	If the volume of water in a solution is increased, what happens to its concentration?	decreases
14	What is the empirical formula?	the simplest ratio of the atoms in a compound
15	How can concentration be calculated in mol/dm^3?	$\dfrac{\text{moles}}{\text{volume}}$
16	What is a titration?	a method used to calculate the concentration of an unknown solution
17	What is the yield of a reaction?	the mass of product you obtained from the reaction
18	What is the end point?	the point at which the reaction is just complete (and no substance is in excess)
19	How should solution be added from the burette close to the end point?	drop by drop (and with swirling)
20	Why is a white tile used in titration?	to better see the colour change
21	What is a titre?	the volume of solution added from the burette
22	What are concordant titres?	two titres within $0.1\ cm^3$ of each other
23	What volume does one mole of any gas occupy at room temperature and pressure?	$24\ dm^3$

C8 Quantitative chemistry

C8

Now use the questions below to check your knowledge from previous chapters.

Previous questions | Answers

#	Question	Answer
1	Why do ionic compounds have high melting points?	strong electrostatic attraction between oppositely charged ions
2	When do ionic compounds conduct electricity?	when molten or in solution
3	How many electrons are in the outer shell of the Group 7 elements?	7
4	What is an ion?	an atom that has lost or gained electrons
5	How are metals more reactive than carbon extracted?	electrolysis
6	Why is gold found as a native metal in the ground?	it is unreactive
7	In the electrolysis of aluminium oxide, why is the aluminium oxide mixed with cryolite?	to lower the melting point
8	What is the test for hydrogen?	a squeaky pop
9	What is the test for oxygen?	relights a glowing splint
10	What is the test for carbon dioxide?	turns limewater milky if bubbled through it

(Put paper here)

Required Practical Skills

Practise answering questions on the required practicals using the example below. You need to be able to apply your skills and knowledge to other practicals too.

Neutralisation reactions	Worked example	Practice
You need to be able to use titration to determine the concentration of a solution of acid or alkali. To do this, you need to be able to describe how to carry out a titration experiment using burettes and pipettes, and how to accurately measure and transfer volumes of liquids. You should also be able to use a volume to calculate the concentration of a solution.	A student carried out a titration, adding sulfuric acid to sodium hydroxide. They repeated their experiment until they had carried out four titres. 1 Describe what concordant results are. Concordant results for a titration are results that are within 0.10 cm³ of each other. A titration should be repeated until you have at least three concordant titres. 2 The student got the following results for the volume of sulfuric acid in cm³ needed to neutralise 25 cm³ sodium hydroxide: 16.05, 16.10, 16.25, 16.00. Calculate the mean volume of acid needed. 16.25 cm³ is not a concordant result so is discarded. $$\text{mean volume} = \frac{(16.05 + 16.10 + 16.00)}{3} = 16.05 \text{ cm}^3$$	1 Describe how to take a reading from a burette. 2 Explain how the end of point of a titration can be determined, naming a suitable indicator and colour changes in your answer.

Practice

Exam-style questions

01 Methane is a compound with the formula CH_4.

01.1 Calculate the relative formula mass M_r of methane. **[1 mark]**
Relative atomic masses A_r: H = 1; C = 12

_____

_____

relative formula mass: _____

> **Exam Tip**
> Show your working clearly.

01.2 Methane reacts with excess oxygen to make carbon dioxide and water.
Methane is the limiting reactant.
What is meant by the term limiting reactant? **[1 mark]**
Tick **one** box.

the reactant present in the smaller mass ☐

the reactant with the smaller relative formula mass ☐

the reactant with the smaller molar mass ☐

the reactant that is completely used up when the other reactant is present in excess ☐

01.3 Write a balanced symbol equation for the reaction between methane and oxygen. **[2 marks]**

_____

01.4 0.13 moles of methane react with 0.25 moles of oxygen. Identify which reactant is the limiting reactant. **[1 mark]**

_____

01.5 Determine how many moles of water will be produced in the reaction in **01.4**. **[1 mark]**

_____

> **Exam Tip**
> You'll need to use the equation for this. If you didn't get the equation correct you can still get some marks by showing your working.

02 This question is about calcium nitrate, $Ca(NO_3)_2$.

02.1 How many oxygen atoms are there in 1 mole of calcium nitrate?
Avogadro constant = 6.02×10^{23} **[2 marks]**

_____

> **Exam Tip**
> Take careful note of which number is inside the brackets and which is outside the brackets.

02.2 Calculate the relative formula mass of calcium nitrate. Use the Periodic Table to help you. **[1 mark]**

Tick **one** box.

102 ☐

150 ☐

164 ☐

204 ☐

02.3 In a fume cupboard, a student heats some calcium nitrate in a test tube.

The calcium nitrate decomposes:

$$2Ca(NO_3)_2(s) \rightarrow 2CaO(s) + 4NO_2(g) + O_2(g)$$

Relative atomic masses A_r: Ca = 40, N = 14, O = 16.

Explain why the mass of solid in the test tube is **lower** after the chemical reaction. **[1 mark]**

_____
_____

> **Exam Tip**
>
> If you see state symbols in a question there is a good chance you need to refer to them in the answer. Look at the changes of state to find the answer to this one.

02.4 In the reaction, 22.4 g of calcium oxide is produced. Calculate the mass of calcium nitrate that reacted. **[5 marks]**

_____
_____
_____
_____

mass = _____ g

03 1 cm³ of water is equal to 1 g of water at room temperature and pressure.

03.1 Determine the volume in cm³ of 1 mole of water. **[2 marks]**

03.2 Calculate the volume in cm³ of one molecule of water. **[2 marks]**

03.3 Use the particle model to explain why the value calculated in **03.2** is not accurate. **[2 marks]**

> **Exam Tip**
>
> Start with working out the M_r of water.

04 Sulfur dioxide reacts with oxygen to make sulfur trioxide.
$$SO_2(g) + 2O_2(g) \rightarrow 2SO_3(g)$$

04.1 Write down what the number 3 means in the formula SO_3. **[1 mark]**

04.2 Calculate the relative formula mass M_r of sulfur dioxide, SO_2. Relative atomic masses A_r: S = 32; O = 16. **[1 mark]**

> **Exam Tip**
> Even for a one-mark question it's important to show your working.

04.3 In an experiment, 1.28 g of sulfur dioxide, SO_2, makes 1.68 g of sulfur trioxide, SO_3. Calculate the mass of oxygen that was needed. **[1 mark]**

05.1 Describe what Avogadro's constant is a measure of. **[1 mark]**

05.2 A glass contains 232 g of water. Estimate the number of water molecules in the glass. Give your answer to three significant figures. Relative atomic masses A_r: H = 1; O = 16. **[5 marks]**

05.3 Deduce the number of water molecules in 464 g of ice. Use your answer to **05.2**. **[2 marks]**

06 Titration can be used to deduce the unknown concentration of an acid. Describe the method for carrying out a titration experiment with sodium hydroxide to determine the concentration of hydrochloric acid. **[6 marks]**

07 Some students investigated the reaction of calcium carbonate with hydrochloric acid:
$$CaCO_3(s) + 2HCl(aq) \rightarrow CaCl_2(aq) + CO_2(g) + H_2O(l)$$

The students measured the volume of carbon dioxide gas made in 60 s. The students repeated the experiment five times. **Table 1** shows their results.

Table 1

Experiment number	Volume of carbon dioxide gas made in 60 s in cm³
1	52
2	49
3	48
4	56
5	55

07.1 Calculate the mean volume of carbon dioxide gas. **[1 mark]**

07.2 Give the range of the values obtained in the five experiments. **[1 mark]**

07.3 Describe how the best estimate of the volume of gas is obtained. Choose **one** answer. **[1 mark]**

mean ± 3   mean ± 4   mean ± 6   mean ± 8

> **Exam Tip**
> You might not have seen numbers written like this before. It is just asking for how high above and below the mean the outermost values are.

C8 Quantitative chemistry

07.4 Hydrochloric acid solution contains hydrogen chloride, HCl, molecules dissolved in water. The students used 25 cm³ of 7.3 g/dm³ hydrochloric acid. Calculate the mass of hydrogen chloride that dissolved. Give your answer to two significant figures. **[3 marks]**

08 A student wants to react nitric acid with potassium hydroxide to form potassium nitrate and water. The balanced symbol for the equation is:

$$HNO_3 + KOH \rightarrow KNO_3 + H_2O$$

08.1 Complete the symbol equation by adding state symbols. **[1 mark]**

08.2 The student dissolved 14 g of potassium hydroxide in 700 cm³ of water. Calculate the concentration of the potassium hydroxide solution in g/dm³. **[2 marks]**

08.3 The concentration of nitric acid was 22 g/dm³. Calculate the mass of nitric acid in 30 cm³ of the solution. **[2 marks]**

08.4 The student reacted 30 cm³ of nitric acid with 35 cm³ of this solution of potassium hydroxide. Identify the limiting reactant. **[6 marks]**

09 Magnesium reacts with nitrogen gas, N_2, to make magnesium nitride, Mg_3N_2.

09.1 Draw a dot and cross diagram to show the bonding in a nitrogen molecule, N_2. **[2 marks]**

> **Exam Tip**
> Don't worry about unfamiliar compounds; you do not need to draw magnesium nitride.

09.2 Two groups of students draw diagrams of apparatus they think could be used to make magnesium nitride from magnesium and nitrogen (**Figure 1**).

Figure 1

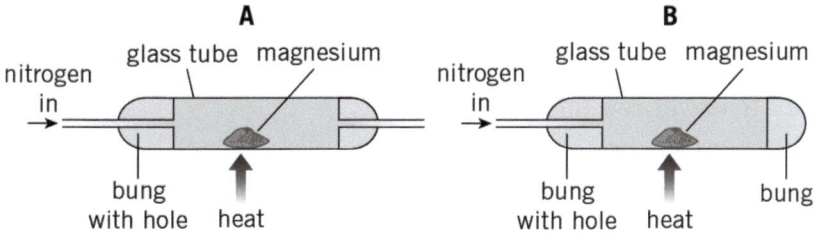

Explain why the apparatus in **B** must **not** be used for the experiment. **[2 marks]**

> **Exam Tip**
> Carefully compare the two images to look for the differences.

09.3 A teacher made magnesium nitride from magnesium and nitrogen. **Table 2** shows the masses of the reactants that reacted and the mass of product made.

Table 2

Substance	Mass in g
magnesium	2.16
nitrogen	0.84
magnesium nitride	3.00

Use the data in **Table 2** to deduce the balanced equation for the reaction. Show all your working and use the data below. Relative atomic masses A_r: Mg = 24; N = 14. **[5 marks]**

> **Exam Tip**
> You can work backwards from the mass.

10 Paracetamol and ibuprofen are painkillers.

10.1 A solution contains 500 mg of paracetamol in 5 cm³ of solution. Calculate the mass of paracetamol in 1 dm³ of solution. Give your answer in g. **[3 marks]**

10.2 The chemical formula of paracetamol is $C_8H_9NO_2$. Calculate the mass of 1 mole of paracetamol. Relative atomic masses A_r: C = 12; H = 1; N = 14; O = 16. **[1 mark]**

> **Exam Tip**
>
> There are lots of parts to this question, write everything down clearly to avoid missing or repeating parts.

10.3 A solution of ibuprofen contains 0.10 g of ibuprofen in 5.0 cm³ of solvent. The chemical formula of ibuprofen is $C_{13}H_{18}O_2$. Calculate the number of moles of ibuprofen in 1 dm³ of the solution. Give your answer to three significant figures. **[4 marks]**

11 Some people take iron tablets if they do not have enough iron in their blood. **Table 3** gives some data about three types of iron tablet.

Table 3

Name of compound	Formula of iron compound in tablet	Mass of iron compound in tablet in g
iron(II) sulfate	$FeSO_4$	0.065
iron(II) fumarate	$C_4H_2FeO_4$	0.076
iron(II) gluconate	$C_{12}H_{24}FeO_{14}$	0.300

11.1 Calculate the relative formula mass of iron(II) fumarate. Relative atomic masses A_r: Fe = 56; C = 12; H = 1; O = 16. **[1 mark]**

> **Exam Tip**
>
> Check the answer to **11.2** before you try **11.3**. They are very similar methods and it's best to correct any mistakes before you move on.

11.2 Calculate the number of moles of iron(II) sulfate in one tablet. Give your answer in standard form to two significant figures. Relative atomic masses A_r: Fe = 56; S = 32; O = 16. **[4 marks]**

11.3 Deduce the mass of iron in one iron(II) gluconate tablet. **[4 marks]**

12 Iron is extracted from its ore in the following reaction. Relative atomic masses A_r: Fe = 56, C = 12, O = 16.

$$2Fe_2O_3 + 3C \rightarrow 4Fe + 3CO_2$$

12.1 Calculate the mass of carbon that reacts with 16.0 g of iron(III) oxide. **[5 marks]**

12.2 Calculate the mass of carbon dioxide produced in the reaction of carbon with 16.0 g of iron(III) oxide. **[4 marks]**

12.3 An industrial plant processes 3.7 tonnes of iron(III) oxide. Calculate the mass in kg of iron produced. 1 tonne = 1000 kg **[6 marks]**

12.4 Calculate the percentage by mass of iron(III) oxide that is iron. **[2 marks]**

13 A student investigated the volume of nitric acid that reacted with 25 cm³ of a solution of sodium hydroxide.

94 C8 Quantitative chemistry

13.1 The student used 25 cm³ of 0.100 mol/dm³ sodium hydroxide. Their results are shown in **Table 4**. Identify the outlier. **[1 mark]**

Table 4

Titration number	Volume of nitric acid solution in cm³
1	13.55
2	12.95
3	13.05
4	13.00

> **Exam Tip**
> Outlier is another term for an anomalous result.

13.2 Calculate the mean volume of nitric acid needed to neutralise 25 cm³ of 0.100 mol/dm³ sodium hydroxide. **[1 mark]**

13.3 Write the balanced symbol equation with state symbols for this reaction. **[3 marks]**

13.4 Calculate the concentration of nitric acid in mol/dm³. Give your answer to two significant figures. **[4 marks]**

> **Exam Tip**
> This is a reaction between nitric acid and sodium hydroxide. You'll need to know the formula of both of these.
> Then use the general salt equation to work out the products, determine the formula of the salt, and balance the equation. This may seem like a lot but you need to learn to apply it all in an exam.

14 A student carried out a titration between hydrochloric acid and sodium hydroxide. This is the method used:

1. Fill a burette with acid.
2. Transfer 25.0 cm³ of sodium hydroxide to a conical flask.
3. Add a few drops of indicator to the sodium hydroxide.
4. Add acid from the burette to the flask until the indicator changes colour.
5. Repeat the procedure.

14.1 Name the piece of apparatus used to transfer the sodium hydroxide in step **2**. **[1 mark]**

14.2 Suggest **two** improvements to step **4**. **[2 marks]**

14.3 Describe how the student will know when to stop repeating the procedure. **[1 mark]**

> **Exam Tip**
> This has to be the best piece of scientific equipment for the job, not anything that can hold liquid.

14.4 **Table 5** shows the results the student obtained.

Table 5

	Titration 1	Titration 2	Titration 3	Titration 4	Titration 5
Initial burette reading in cm³	1.20	24.50	0.80	21.10	20.50
Final burette reading in cm³	24.50	45.70	21.10	42.35	40.75
Volume of acid added in cm³	23.30	21.20	21.30	21.25	20.25

Use the student's results to calculate the mean volume of acid added. Give your answer to one decimal place. **[1 mark]**

14.5 The equation for the reaction is:

$$NaOH(aq) + HCl(aq) \rightarrow NaCl(aq) + H_2O(l)$$

The concentration of acid used was 0.200 mol/dm³. Calculate the concentration of sodium hydroxide in mol/dm³. Give your answer to three significant figures. **[4 marks]**

15 A student wanted to make large copper sulfate crystals from copper hydroxide and an acid.

15.1 Name the acid the student should use. **[1 mark]**

> **Exam Tip**
> The second part of the salt name should point you towards the acid used.

15.2 Describe how the student could make a sample of copper sulfate crystals from copper hydroxide and the acid. In your answer:
- Name the pieces of apparatus required.
- Give a reason for each step. **[6 marks]**

15.3 Write a balanced symbol equation with state symbols for the reaction between the acid and copper hydroxide, $Cu(OH)_2$. **[3 marks]**

> **Exam Tip**
> Don't forget to add in state symbols. They are often forgotten but have been mentioned in the question, so there will be separate marks for them. Remember, if something is dissolved in water it is aqueous (aq), but water itself is a liquid (l).

15.4 The student used 30 cm³ of 32.5 g/dm³ copper hydroxide solution. Calculate the moles of copper hydroxide that were used in the reaction. $M_r(Cu(OH)_2) = 97.5$ **[4 marks]**

15.5 Calculate the maximum mass of copper sulfate that the student could produce. Give your answer to two significant figures. $M_r(CuSO_4) = 159.5$ **[5 marks]**

16 A student made some zinc nitrate crystals by reacting zinc carbonate with nitric acid. The equation for the reaction is:

$$ZnCO_3(s) + 2HNO_3(aq) \rightarrow Zn(NO_3)_2(aq) + CO_2(g) + H_2O(l)$$

The mass of zinc nitrate made was 9.45 g.

16.1 Calculate the mass of zinc carbonate that reacted to make 9.45 g of zinc nitrate. **[5 marks]**

> **Exam Tip**
> Take this question part by part – pH before, pH as it was added, and then pH at the end.

16.2 To make the zinc nitrate, the student started with some acid in a conical flask. The student then added zinc carbonate, a little at a time, until some remained unreacted. Explain how the pH of the mixture in the conical flask changed as the student added zinc carbonate. Include a possible pH value for the acid at the start of the reaction. **[3 marks]**

17 Some students carried out a titration between sodium hydroxide and sulfuric acid. They used the following method:
1. Use a measuring cylinder to transfer 25.0 cm³ of sodium hydroxide to a conical flask.
2. Fill a burette with acid.
3. Add 1 cm³ of indicator to the flask.
4. Add acid from the burette to the flask until the indicator changes colour.
5. Repeat the procedure, adding the acid drop by drop as the end point approaches.

96 C8 Quantitative chemistry

17.1 Suggest an improvement to step **1**. [1 mark]

> **Exam Tip**
> Think of a more precise piece of equipment you could use.

17.2 Suggest **two** things the students can do to avoid spillages in step **2**. [2 marks]

17.3 Identify the mistake in step **3**. [1 mark]

17.4 **Table 6** shows their results.

Table 6

Titration	Initial burette reading in cm³	Final burette reading in cm³	Volume of acid added in cm³
1	1.30	18.35	19.65
2	19.65	37.85	18.20
3	2.40	20.70	18.30
4	20.70	39.95	
5	0.05	18.30	18.25

Write down the missing volume in **Table 6**. [1 mark]

17.5 Use the students' results to calculate the mean volume of acid added. [1 mark]

> **Exam Tip**
> Only use the concordant results.

17.6 The equation for the reaction is:
$$H_2SO_4(aq) + 2NaOH(aq) \rightarrow Na_2SO_4(aq) + 2H_2O(l)$$
The concentration of acid used was 0.100 mol/dm³. Calculate the concentration of sodium hydroxide in mol/dm³. Give your answer to three significant figures. [5 marks]

18 A student has a solution of 20 g/dm³ sodium hydroxide and sulfuric acid of an unknown concentration.

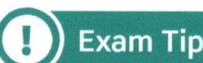

18.1 Write a balanced symbol equation for the reaction between sodium hydroxide and sulfuric acid. Include state symbols. [3 marks]

> **Exam Tip**
> Getting the ratio of ions correct in sodium sulfate is a topic students often make mistakes on!

18.2 Describe a titration method by which the student can determine the concentration of sulfuric acid. [6 marks]

18.3 The student determines that the concentration of sulfuric acid was 29.4 g/dm³. The student used 25 cm³ of sulfuric acid in their titration. Calculate the number of moles of sodium hydroxide that was needed to neutralise 25 cm³ of sulfuric acid. [6 marks]

> **Exam Tip**
> There are lots of numbers in this question. Make a list of the key ones here so you can easily refer back to it and don't get the numbers confused.

18.4 Calculate the volume of sodium hydroxide that was required to neutralise 25 cm³ of sulfuric acid. [6 marks]

19 Lead nitrate solution reacts with sodium iodide solution to make lead iodide, PbI_2, and sodium nitrate, $NaNO_3$. The lead iodide forms as a precipitate. Sodium nitrate is in solution.

19.1 Define the law of the conservation of mass. [2 marks]

19.2 Lead nitrate reacts with sodium iodide. Write a balanced equation, including state symbols, for the reaction. The formula of lead nitrate is $Pb(NO_3)_2$. **[3 marks]**

Exam Tip

Determine the relative formula mass of each substance.

You will need to use your symbol equation from **19.2** to answer **19.3**.

19.3 Calculate the minimum mass of sodium iodide required to make 6.68 g of lead iodide. Give your answer to two significant figures.
Relative atomic mass A_r: Na = 23; Pb = 207; I = 127 **[5 marks]**

20 This question is about reactions of magnesium.

20.1 Which of these metals has the greatest tendency to form positive ions? Choose **one** answer. **[1 mark]**

iron   lithium   magnesium   zinc

20.2 Name the product formed in the reaction between magnesium and oxygen. **[1 mark]**

20.3 Identify whether magnesium is oxidised or reduced in the reaction in **20.2**. Give a reason for your answer. **[2 marks]**

20.4 Magnesium cannot be extracted from the compound formed in **20.2** by reaction with carbon. Explain why. **[2 marks]**

21 This question is about the pH scale. A student measured the pH of some solutions. **Table 7** shows the results the student obtained.

Table 7

Solution	pH
A	7
B	2
C	10
D	5
E	12

21.1 Name **two** ways of measuring the pH of a solution. **[2 marks]**

21.2 Give the letter of the neutral solution. **[1 mark]**

21.3 Give the letter of the most alkaline solution. **[1 mark]**

21.4 Give the letter of the solution that has the highest concentration of hydrogen ions, H^+. **[1 mark]**

Exam Tip

Remember, pH is a measure of the number of H^+ ions in a solution.

21.5 Some alkali is added to solution **A**. Write whether the pH increases, decreases, or stays the same. **[1 mark]**

22.1 Draw **one** line from each state symbol to the correct state it represents, and to the correct particle diagram for that state.
[3 marks]

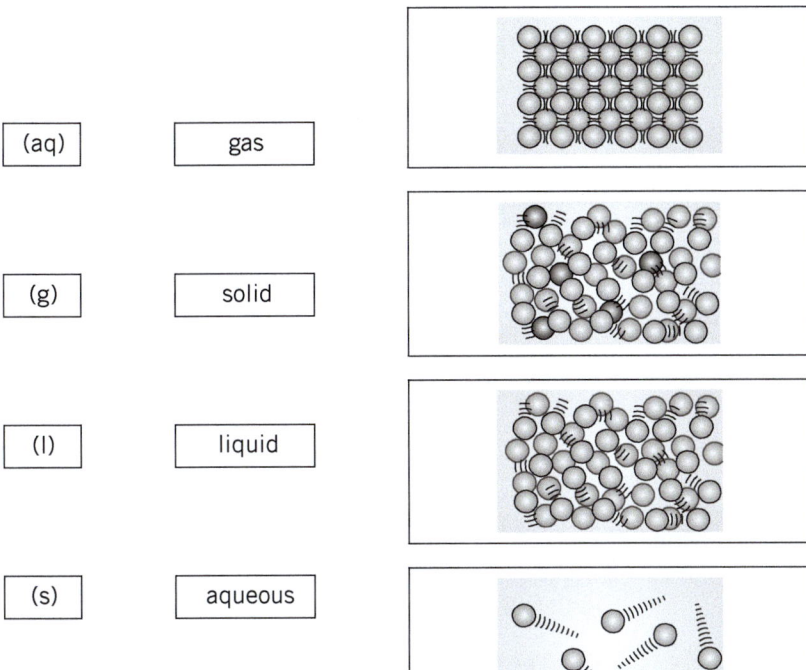

(aq)	gas
(g)	solid
(l)	liquid
(s)	aqueous

22.2 A student added solid magnesium to hydrochloric acid. The reaction produced hydrogen gas and magnesium chloride. The magnesium chloride is dissolved in water. Complete the symbol equation for the reaction by adding state symbols. **[2 marks]**

Mg ____ + 2HCl ____ → H$_2$ ____ + MgCl$_2$ ____

22.3 Calculate the relative formula mass of magnesium chloride. Choose **one** answer. **[1 mark]**

29   46   59.5   95

22.4 The concentration of hydrochloric acid used in the reaction was 19 g/dm³. Calculate the mass of hydrochloric acid in 0.5 dm³.
[1 mark]

! Exam Tip
Make sure you're using the correct numbers from the Periodic Table. You need to use the mass number, not the atomic number.

22.5 The student carried out the reaction in a conical flask placed on a top-pan balance. Complete the sentences. Choose the words from the box.

| decreased | increased | gas | liquid | solution |

The student observed that as the reaction proceeded the mass _____. This is because a _____ was produced.

! Exam Tip
One of the products will escape from the conical flask.

Knowledge

C9 Rates of reaction

Rates of reaction

The **rate of a reaction** is how quickly the reactants turn into the products.

To calculate the rate of a reaction, you can measure:

- how quickly a reactant is used up

$$\text{mean rate of reaction} = \frac{\text{quantity of reactant used}}{\text{time taken}}$$

- how quickly a product is produced.

$$\text{mean rate of reaction} = \frac{\text{quantity of product formed}}{\text{time taken}}$$

For reactions that involve a gas, this can be done by measuring how the mass of the reaction changes or the volume of gas given off by the reaction.

Volume of gas produced

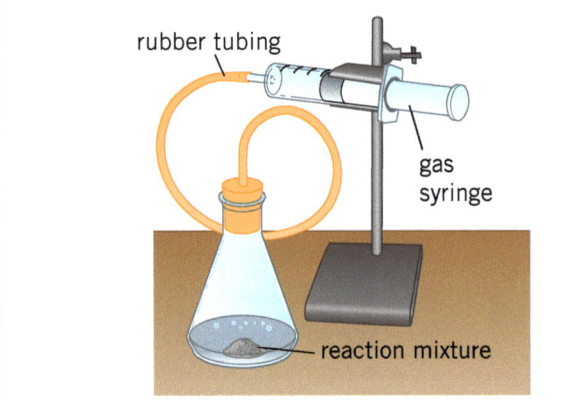

The reaction mixture is connected to a gas syringe or an upside down measuring cylinder. As the reaction proceeds the gas is collected.

The rate for the reaction is then:

$$\text{rate} = \frac{\text{volume of gas produced}}{\text{time taken}}$$

Volume is measured in cm³ and time in seconds, so the unit for rate is cm³/s.

Calculating rate from graphs

The results from an experiment can be plotted on a graph.

- A steep **gradient** means a high rate of reaction – the reaction happens quickly.
- A shallow gradient means a low rate of reaction – the reaction happens slowly.

Change in mass

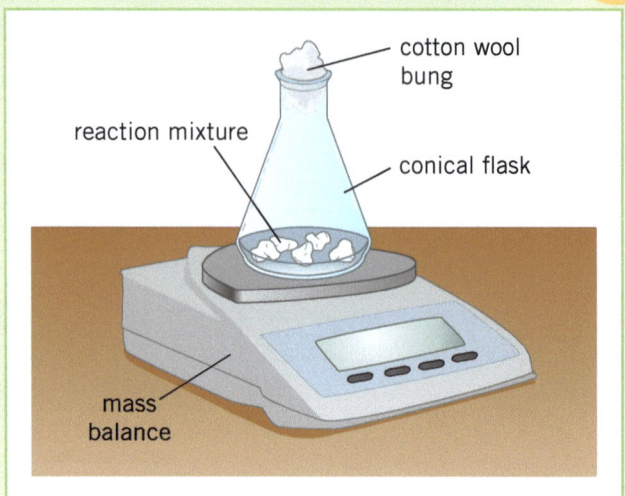

The reaction mixture is placed on a mass balance. As the reaction proceeds and the gaseous product is given off, the mass of the flask will decrease.

The rate for the reaction is then:

$$\text{rate} = \frac{\text{change in the mass}}{\text{time taken}}$$

The mass is measured in grams and time is measured in seconds. Therefore, the unit of rate is g/s.

Mean rate between two points in time

To get the **mean rate** of reaction between two points in time:

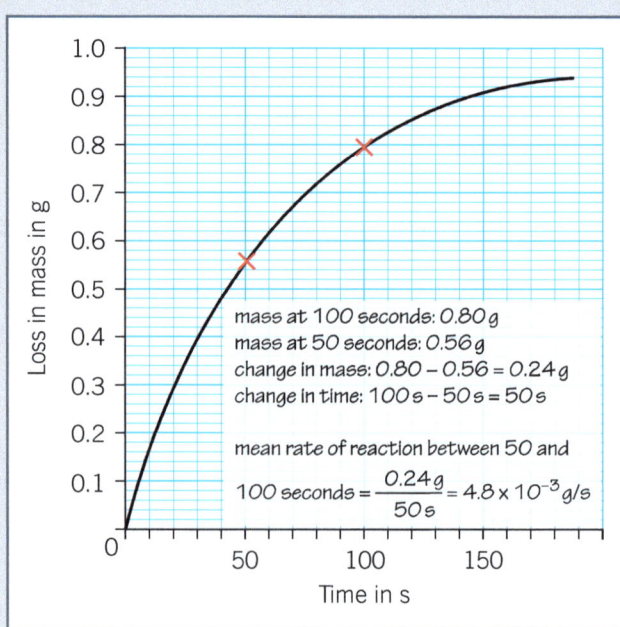

mass at 100 seconds: 0.80 g
mass at 50 seconds: 0.56 g
change in mass: 0.80 − 0.56 = 0.24 g
change in time: 100 s − 50 s = 50 s

mean rate of reaction between 50 and

$$100 \text{ seconds} = \frac{0.24 \text{ g}}{50 \text{ s}} = 4.8 \times 10^{-3} \text{ g/s}$$

C9

Collision theory

For a reaction to occur, the reactant particles need to collide. When the particles collide, they need to have enough energy to react or they will just bounce apart. This amount of energy is called the **activation energy**.

You can increase the rate of a reaction by:

- increasing the **frequency of collisions**
- increasing the energy of the particles when they collide.

Factors affecting rate of reaction

Condition that increases rate	How is this condition caused?	Why it has that effect
increasing the temperature	heat the container in which the reaction is taking place	1 particles move faster, leading to more frequent collisions 2 particles have more energy, so more collisions result in a reaction note that these are two *separate* effects
increasing the concentration of solutions	use a solution with more solute in the same volume of solvent	there are more reactant particles in the reaction mixture, so collisions become more frequent
increasing the pressure of gases	increase the number of gas particles you have in the container or make the container smaller	less space between particles means more frequent collisions
increasing the surface area of solids	cut the solid into smaller pieces, or grind it to create a powder, increasing the surface area. Larger pieces decrease the surface area	only reactant particles on the surface of a solid are able to collide and react; the greater the surface area the more reactant particles are exposed, leading to more frequent collisions

Catalysts

Some reactions have specific substances called catalysts that can be added to increase the rate. These substances are not used up in the reaction.

A catalyst provides a different reaction pathway that has a lower activation energy. As such, more particles will collide with enough energy to react, so more collisions result in a reaction.

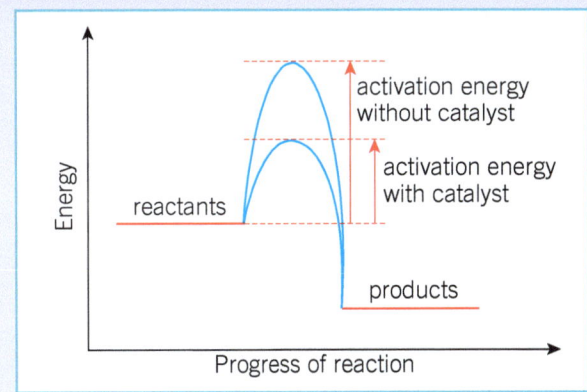

Key Terms

Make sure you can write a definition for these key terms.

activation energy   collision   collision theory   frequency of collision

gradient   mean rate   rate of reaction

Retrieval

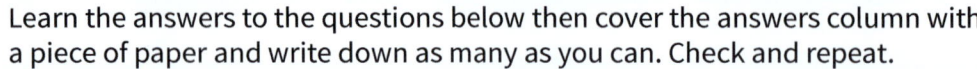

Learn the answers to the questions below then cover the answers column with a piece of paper and write down as many as you can. Check and repeat.

C9 questions | Answers

#	Question	Answer
1	What is the rate of a reaction?	how quickly it occurs
2	What is the equation for calculating rate of reaction from a change in mass?	$\dfrac{\text{change in mass}}{\text{time}}$
3	What is the unit for rate of reaction in a reaction involving a change in mass?	g/s
4	What is the equation for calculating rate of reaction from a change in volume?	$\dfrac{\text{change in volume}}{\text{time}}$
5	What is the unit for rate of reaction in a reaction involving a change in volume?	cm^3/s
6	If a rate of reaction curve has a steep gradient at a certain point, what does that tell you about the rate?	it is high
7	If a rate of reaction curve has a shallow gradient at a certain point, what does that tell you about the rate?	it is low
8	What is collision theory?	that reactants need to collide in order to react
9	What is the activation energy?	the minimum amount of energy colliding particles have to have before a reaction will take place
10	Describe the effect of increasing concentration on the rate of reaction.	increases
11	Explain the effect of increasing concentration on the rate of reaction.	more reactant particles in the same volume lead to more frequent collisions
12	Describe the effect of increasing the pressure on the rate of reaction.	increases
13	Explain the effect of increasing the pressure on the rate of reaction.	less space between particles means more frequent collisions
14	Describe the effect of increasing the surface area on the rate of reaction.	increases
15	Explain the effect of increasing the surface area on the rate of reaction.	more reactant particles are exposed and able to collide, leading to more frequent collisions
16	Describe the effect of increasing the temperature on the rate of reaction.	increases
17	Explain the effect of increasing the temperature on the rate of reaction.	two separate effects: particles move faster, leading to more frequent collisions; particles have more energy, so more collisions result in a reaction
18	What is a catalyst?	a substance that increases the rate of a reaction but is not used up in the reaction
19	How do catalysts increase the rate of a reaction?	lower the activation energy of the reaction, so more collisions result in a reaction
20	Why is the rate of a reaction highest at the start of a reaction?	there are lots of reactant particles so frequent collisions
21	Why does the rate of a reaction decrease over time?	fewer reactant particles, less frequent collisions

C9

Now use the questions below to check your knowledge from previous chapters.

Previous questions | Answers

#	Question	Answer
1	How many electrons go in the first shell?	2
2	How many electrons go in the second and third shell?	8
3	Describe the structure and bonding in graphite.	each carbon atom is bonded to three others in hexagonal rings arranged in layers. It has delocalised electrons and no bonds between the layers
4	Explain why graphite can conduct electricity.	the delocalised electrons can move through the graphite
5	Which is the least reactive alkali metal.	lithium
6	Name two ways of extracting ores?	phytomining and bioleaching
7	What are the three products of the electrolysis of sodium chloride solution?	hydrogen, sodium hydroxide, chlorine
8	What are the reasons for electroplating a metal?	increase durability, improve desirability, reduce corrosion
9	What is the solvent front?	the point which the solvent reaches up to on the stationary phase
10	How is R_f calculated?	distance moved by spot/solvent front
11	When a metal forms a metal oxide, why does the mass increase?	because oxygen atoms have been added
12	When an acid reacts with a metal, why does the mass decrease?	because a gas is produced and escapes

Put paper here

Practical Skills

Practise answering questions on the required practicals using the example below. You need to be able to apply your skills and knowledge to other practicals too.

Rates of reaction	Worked example	Practice
From this practical, you should be able to find the rate of a reaction: 1 Measure the production of a gas. 2 Measure changes in the colour or turbidity of a solution.	Silver chloride is an insoluble salt that can be made in the following reaction. Suggest a how the rate of this reaction could be measured. $AgNO_3(aq) + NaCl(aq) \rightarrow NaNO_3(aq) + AgCl(s)$ **Answer:** The reactants are both colourless solutions. Solid silver chloride will form as a precipitate and make the solution appear cloudy. One way of measuring the rate of the reaction is to look at the rate of production of silver chloride precipitate. This could be measured by placing the beaker with the reacting solution on a piece of white paper with a black cross printed on it, and measuring the time taken for the cross to disappear.	1 Give three factors that can affect the rate of a reaction. 2 Give two methods that can be used to determine the rate of a reaction where a gas is produced. 3 Suggest another method to measure the rate of the production of silver chloride precipitate.

Practice

Exam-style questions

01 Potassium iodide acts as a catalyst for the decomposition of hydrogen peroxide.

01.1 Describe what a catalyst does. **[1 mark]**

Tick **one** box.

Decreases reaction rate by providing a pathway with a higher activation energy. ☐

Decreases reaction rate by providing a pathway with a lower activation energy. ☐

Increases reaction rate by providing a pathway with a higher activation energy. ☐

Increases reaction rate by providing a pathway with a lower activation energy. ☐

> **Exam Tip**
>
> The are two parts to **01.1**. The first part relates to reaction rate and the second part relates to activation energy. Decide how a catalyst affects reaction rate and cross off the two wrong answers. Then decide how a catalyst affects the activation energy and cross off the one remaining wrong answer.

01.2 **Figure 1** shows the reaction profile for the catalysed reaction.

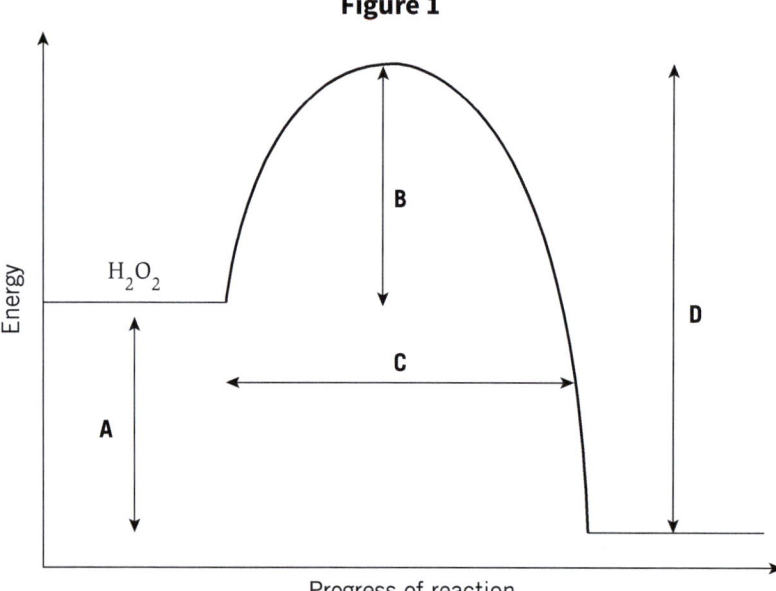

Figure 1

Identify the arrow that shows the activation energy for the catalysed reaction. **[1 mark]**

_____

01.3 Balance the symbol equation for the decomposition of hydrogen peroxide. Give the missing state symbol. **[2 marks]**

____$H_2O_2(aq)$ → ____$H_2O($ ____ $) + O_2(g)$

01.4 Explain why potassium iodide is not given in the balanced symbol equation. **[1 mark]**

_____

C9 Rate of reaction

02 Magnesium reacts with dilute hydrochloric acid:

$$Mg + 2HCl \rightarrow MgCl_2 + H_2$$

A student investigated how the volume of hydrogen produced changed over time.

Table 1 shows the student's results.

Table 1

Time in seconds	Total volume of hydrogen produced in cm³
0	0
30	22
60	38
90	52
120	58
150	61
180	61

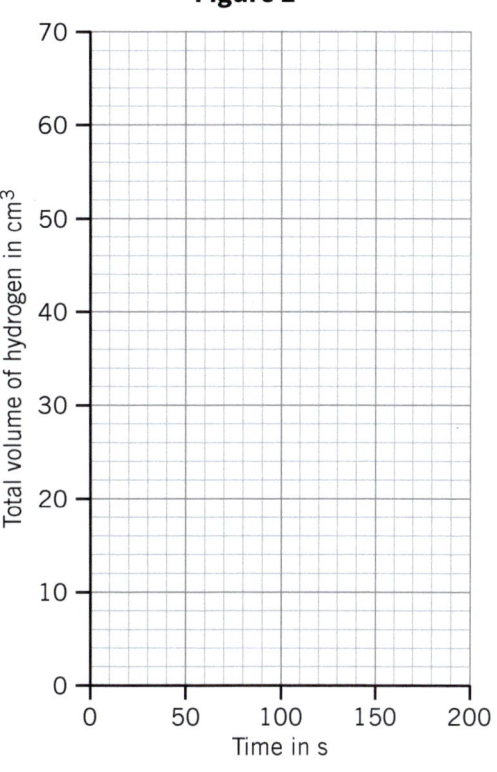

Figure 2

02.1 Plot the data on **Figure 2**. [2 marks]

02.2 Draw a line of best fit. [1 mark]

02.3 Calculate the rate of reaction at 100 seconds.
Give your answer to two significant figures. [4 marks]

> **Exam Tip**
> It might help you if you mark 30, 60, 90, etc. on the x-axis to help you accurately plot the points.

rate at 100 seconds = _____ cm³/s

02.4 The rate of the reaction at 10 seconds is 0.83 cm³/s.
Suggest a reason for the difference in rate at 10 seconds and 100 seconds. [1 mark]

_____

_____

03 Sodium thiosulfate solution reacts with hydrochloric acid. One of the products of the reaction is sulfur, which forms as a precipitate.
$Na_2S_2O_3(aq) + 2HCl(aq) \rightarrow 2NaCl(aq) + H_2O(l) + SO_2(g) + S(s)$

Some students investigated the rate of reaction at different temperatures. This is the method used:

1. Place 50 cm³ of sodium thiosulfate in a conical flask.
2. Start a timer then add 5 cm³ of hydrochloric acid to the flask.
3. Look down through the flask at a cross drawn on a piece of paper.
4. Stop the timer when the cross disappears.
5. Repeat the experiment, each time heating the sodium thiosulfate to a different temperature.

03.1 Suggest **one** improvement that could be made to step **2** to ensure that the results at different temperatures are comparable. **[1 mark]**

03.2 One student suggests measuring the temperature of the reaction mixture after adding the hydrochloric acid instead of measuring the temperature of the sodium thiosulfate on its own. Suggest an advantage of this idea. **[1 mark]**

03.3 Suggest why the same student in the group should carry out step **4** at every temperature. **[1 mark]**

03.4 **Table 2** shows the students' results. Identify which result is anomalous. **[1 mark]**

03.5 Describe how the rate of reaction changes between 0 °C and 61 °C. **[1 mark]**

03.6 Explain why the rate of reaction changes between 0 °C and 61 °C. **[2 marks]**

Table 2

Temperature in °C	Time for X to disappear in seconds
0	180
21	44
39	22
45	21
52	9
61	5

> **Exam Tip**
>
> It might seem odd that **03.5** and **03.6** are nearly the same question but the command word tells you they are looking for different answers. *Describe* is what it looks like, and *explain* is the why.

04 Calcium carbonate reacts with hydrochloric acid:
$2HCl(aq) + CaCO_3(s) \rightarrow CaCl_2(aq) + CO_2(g) + H_2O(l)$

Some students want to investigate how the size of the pieces of solid calcium carbonate affects the rate of the reaction. **Figure 3** shows the apparatus.

Figure 3

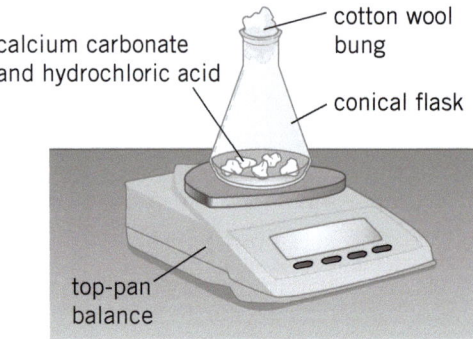

106 C9 Rate of reaction

This is the method used:

1. Weigh approximately 3.0 g of calcium carbonate onto a piece of paper and leave it all on the balance.
2. Place 100 cm³ of dilute hydrochloric acid in a conical flask and place on the balance.
3. Zero the balance.
4. Add the calcium carbonate to the acid and start the stopwatch.
5. Leave the flask and its contents on the balance but remove the paper.
6. Record the time for the total mass to decrease by 0.50 g.
7. Repeat with different sized pieces of calcium carbonate.

04.1 Explain why the total mass of the contents of the conical flask decreases. **[1 mark]**

> **Exam Tip**
> Look at the state symbols of the products.

04.2 Suggest an improvement to step **5**. Give a reason for your answer. **[2 marks]**

04.3 **Table 3** shows the students' results.

Table 3

Size of calcium carbonate pieces	Time for mass to decrease by 0.50 g in seconds
large	1280
medium	690
small	302

Explain the pattern shown in **Table 3**. **[3 marks]**

> **Exam Tip**
> The students have used the same mass of calcium carbonate (as stated in the method). What will be the difference between one large piece of calcium carbonate and lots of small pieces of calcium carbonate?

05 Zinc is a metal. It reacts with dilute nitric acid. The products of the reaction are zinc nitrate and a gas.

05.1 Name the gas formed in the reaction. **[1 mark]**

05.2 Identify which one of these changes will make the reaction rate slower. Choose **one** answer. **[1 mark]**

decreasing the acid concentration
increasing the pressure
decreasing the size of the pieces of zinc
increasing the temperature

> **Exam Tip**
> **05.2** is asking about temperature and rate in the opposite way to how it is normally asked. The chemistry is still the same so just apply your knowledge to the slightly different situation.

05.3 Predict the effect of decreasing the temperature on the rate of reaction. **[1 mark]**

06 Some students investigated the factors that affect the rate of the reaction of magnesium with excess hydrochloric acid. They followed the reaction by measuring the total volume of gas formed every 30 seconds. They changed the conditions and repeated the experiment. **Figure 4** is a graph of some of the results of the two experiments, **P** and **Q**. Curve **P** shows the results for the first experiment.

Exam Tip

Both reactions end at the same point, that's why the graphs level off at the same point for both lines. The rate will only affect how *quickly* a reaction finishes, not how much product is produced.

Figure 4

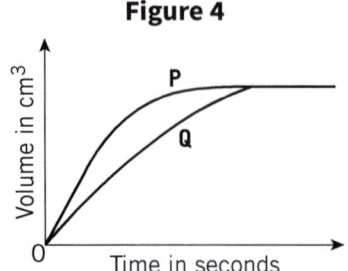

06.1 Write the balanced symbol equation with state symbols for the reaction between magnesium and hydrochloric acid. **[3 marks]**

06.2 Curve **Q** was obtained in the second experiment. Suggest **one** variable that the students might have changed and **two** variables that they kept constant in order to obtain curve **Q**. Justify your answer. **[6 marks]**

07 A student investigated the catalytic decomposition of hydrogen peroxide:

$$2H_2O_2 \,(aq) \rightarrow 2H_2O(l) + O_2(g)$$

Figure 5 shows the apparatus used.

Figure 5

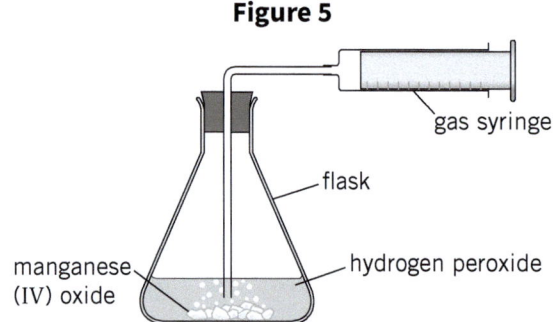

07.1 The student made a mistake in setting up the apparatus. Describe how the student must improve the apparatus before doing the experiment. Give a reason for making this improvement. **[2 marks]**

07.2 The student improved the apparatus set-up and collected some data. **Table 4** shows the results the student obtained.

Calculate the average rate of reaction. Give the unit of the rate of the reaction.

[3 marks]

Table 4

Time in minutes	Volume of gas produced in cm³
0	0
1	42
2	69
3	86
4	88
5	91
6	91

Exam Tip

The number of significant figures in your answer should match the data provided.

Exam Tip

Use the units of the variables to help you work out the units for the rate.

07.3 Predict how the mean rate of reaction would change if powdered manganese(IV) oxide was used instead of lumps.
Give a reason for your prediction. **[2 marks]**

08 A student investigated the decomposition of magnesium carbonate.
$$MgCO_3(s) \rightarrow MgO(s) + CO_2(g)$$

The student measured the volume of carbon dioxide produced.
Table 5 shows their results.

Table 5

Time in s	0	20	40	60	80	100	120	140	160
Volume of carbon dioxide produced in cm³	0	20	39	55	64	65	73	75	75

08.1 Plot the student's results on **Figure 6**. Draw a line of best fit. **[3 marks]**

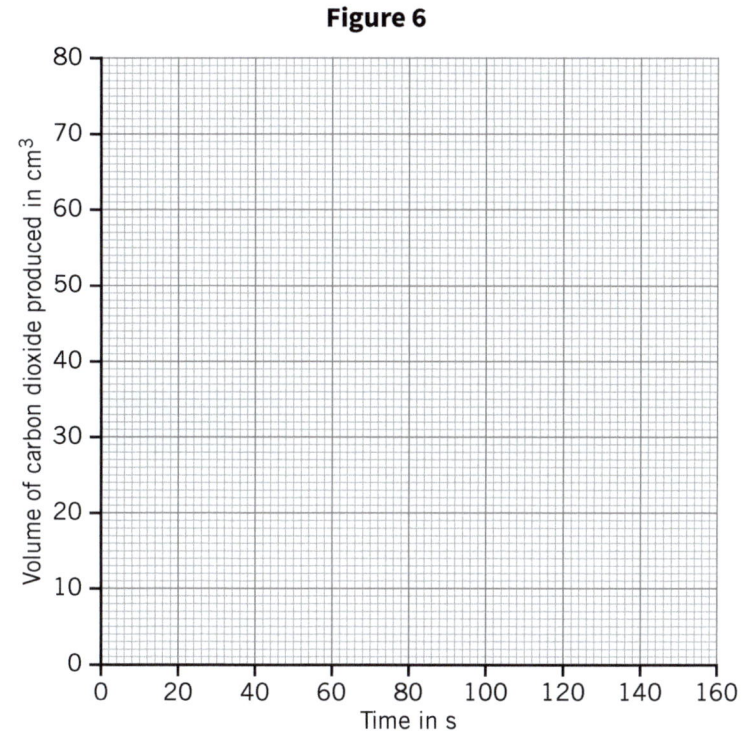

Figure 6

> **Exam Tip**
> Your points must be plotted using a sharp pencil, a cross placed directly over the point you're indicating. Dots or circles are not clear enough as they can cover many points.

> **Exam Tip**
> Your line of best fit may be straight or curved and must go through the majority of the points, but not always all of them. Draw it with a smooth single line.

08.2 Identify the anomalous point. **[1 mark]**

08.3 Explain, in terms of particles, why the rate of reaction was faster at the start of the reaction than towards the end of the reaction. **[4 marks]**

09 A student investigates the effect of concentration on the rate of reaction between nitric acid and sodium carbonate.

09.1 Write a method the student could use. Your method should include how you will measure the rate of reaction and the variables you will control. **[6 marks]**

09.2 Write a prediction for the student's investigation. Explain your prediction. **[3 marks]**

09.3 Another student investigated the effect of the surface area of sodium carbonate on the rate of the reaction. Their results are shown in **Table 6**.

Table 6

Surface area of sodium carbonate	Time taken to produce 500 cm³ of carbon dioxide in s	Mean rate of reaction in _____
solid pieces	195	2.7
powder	42	

Give the unit of the rate of reaction. **[1 mark]**

09.4 Calculate the mean rate of reaction for powdered sodium carbonate. Give your answer to three significant figures. **[2 marks]**

09.5 Give **one** other factor that will affect the rate of the reaction between nitric acid and sodium carbonate. **[1 mark]**

10 Sodium thiosulfate reacts with hydrochloric acid in the following reaction:
$$Na_2S_2O_3(aq) + 2HCl(aq) \rightarrow 2NaCl(aq) + H_2O(l) + SO_2(g) + S(s)$$

10.1 Explain how increasing the concentration of hydrochloric acid will affect the rate of the reaction. **[1 mark]**

10.2 Explain how increasing the temperature will affect the rate of the reaction. **[4 marks]**

10.3 The rate of the reaction between sodium thiosulfate and hydrochloric acid can be determined by measuring how long it takes for the reaction mixture to become cloudy. A student uses a light sensor and data logger to measure the turbidity of the reaction. Sketch a graph of time against turbidity to show the rate of the reaction at two different temperatures. **[4 marks]**

> **Exam Tip**
> Sketching graphs doesn't mean plotting points, just the labelled axis and the line are all that's needed.

11 A teacher demonstrated the thermite reaction. In the thermite reaction, iron(III) oxide reacts with aluminium. Large amounts of heat are transferred to the surroundings. The equation for the reaction is:
$$Fe_2O_3 + 2Al \rightarrow 2Fe + Al_2O_3$$
Relative atomic masses A_r: Fe = 56; O = 16; Al = 27

11.1 Suggest **two** safety precautions that the teacher should take. **[2 marks]**

> **Exam Tip**
> Don't worry about the (III) in iron(III) oxide, it's just part of the compound name.

11.2 The teacher used 8.0 g of iron(III) oxide and 2.7 g of aluminium. Show that neither reactant is present in excess. **[4 marks]**

11.3 Calculate the maximum theoretical yield of iron made from 8.0 g of iron(III) oxide. **[2 marks]**

11.4 Suggest **two** reasons why the percentage yield of the thermite reaction is not 100%. **[2 marks]**

> **Exam Tip**
> All of the information given to you in the question needs to be used, including in the main body of the question text.

C9

12 Sulfuric acid reacts with sodium hydroxide solution:

$$H_2SO_4 + 2NaOH \rightarrow Na_2SO_4 + 2H_2O$$

In an experiment, 25.0 cm³ of 0.100 mol/dm³ sodium hydroxide solution reacts with 27.5 cm³ of sulfuric acid.

12.1 Calculate the concentration of the sulfuric acid in mol/dm³. Give your answer to three significant figures. **[5 marks]**

12.2 Calculate the concentration of the sulfuric acid in g/dm³. Use your answer to **12.1**. Relative atomic masses A_r: H = 1; S = 32; O = 16 **[2 marks]**

12.3 Calculate the mass of sodium hydroxide in 1 dm³ of 0.100 mol/dm³ solution. Relative atomic masses A_r: Na = 23; O = 16; H = 1 **[2 marks]**

> **Exam Tip**
>
> Pick out all the key bits of data you need for this and keep them in one place. This will stop you needing to back over the text.
>
> Write down the volume of acid, volume of alkali, concentration of alkali, and ratio of acid to alkali.

13 **Figure 7** is an outline of the Periodic Table.

Each element is represented by a letter. The letters are **not** the chemical symbols of the elements.

Figure 7

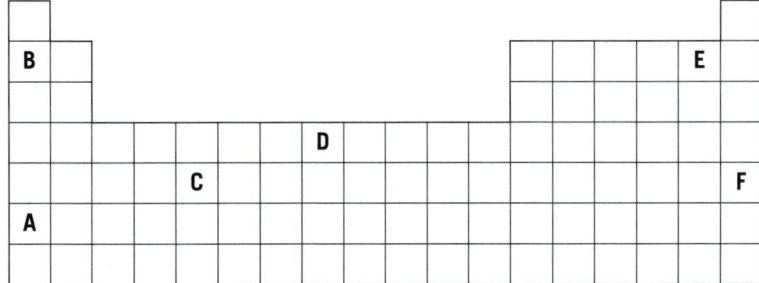

> **Exam Tip**
>
> Draw a line to divide metals and non-metals. This will help with these questions.

13.1 Give the letters of **two** elements that do not conduct electricity. **[1 mark]**

13.2 Give the letters of the **two** elements that react together most vigorously. **[1 mark]**

13.3 Compare the physical and chemical properties of elements **B** and **D**. **[6 marks]**

14 **Table 7** shows data for some elements.

Table 7

Element	Melting point in °C	Relative conductivity
caesium	29	0.53
copper	1083	5.90
gold	1063	4.20
iron	1535	1.00
lithium	180	1.10
sodium	98	2.20

Transition metals have higher melting points and conductivities than Group 1 elements. Evaluate the statement above using the data in **Table 7** only. **[6 marks]**

> **Exam Tip**
>
> The question says "using the data in **Table 7** only", meaning you won't gain any marks for information you write that isn't in the table or backed up by data from the table.

C9 Practice 111

Knowledge

C10 The extent of reactions

Reversible reactions

In some reactions, the products can react to produce the original reactants. This is called a **reversible reaction**. When writing chemical equations for reversible reactions, use the ⇌ symbol.

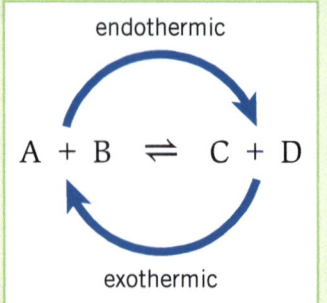

In this reaction:

- A and B can react to form C and D – the forward reaction
- C and D can react to form A and B – the reverse reaction.

The different directions of the reaction have opposite energy changes.

If the forward reaction is *endothermic*, the reverse reaction will be *exothermic*.

The same amount of energy is transferred in each direction.

Equilibrium

In a **closed system** no reactants or products can escape. If a reversible reaction is carried out in a closed system, it will eventually reach **dynamic equilibrium** – a point in time when the forward and reverse reactions have the same rate.

At dynamic equilibrium:

- the reactants are still turning into the products
- the products are still turning back into the reactants
- the rates of these two processes are *equal*, so overall the amount of reactants and products are constant.

Dynamic equilibrium

At dynamic equilibrium the amount of reactant and product are constant, but not necessarily equal.

You could have a mixture of reactants and products in a 50:50 ratio, in a 75:25 ratio, or in any ratio at all. The **conditions** of the reaction are what change that ratio.

How dynamic equilibrium is reached

Progress of reaction	start of reaction	middle of reaction	at dynamic equilibrium
Amount of A + B	high	decreasing	constant
Frequency of collisions A + B	high	decreasing	constant
Rate of forward reaction	high	decreasing	same as rate of reverse reaction
Amount of C + D	zero	increasing	constant
Frequency of collisions C + D	no collisions	increasing	constant
Rate of reverse reaction	zero	increasing	same as rate of forward reaction

Reaction conditions

The conditions of a reaction refer to the external environment of the reaction. When the reaction occurs in a closed system, you can change the conditions by:

- changing the concentration of one of the substances
- changing the temperature of the entire reaction vessel
- changing the pressure inside the vessel.

At equilibrium, the amount of reactants and products is constant. In order to change the amounts the *conditions* of the reaction must be changed, which will favour either the forward or the reverse reaction.

For example, lowering the concentration of the product in the system causes the forward reaction to be **favoured** to increase the concentration of the product. Increasing the temperature of the reaction favours the endothermic direction. When the reaction is in the gas phase, increasing the pressure will favour the side with the fewest total moles of gas.

C10

The Haber process

Fertilisers are important chemicals used to improve the growth of crop plants. Ammonia is a vital component of many fertilisers. It is produced in the **Haber process**. It is a reversible reaction, so the conditions affect the yield.

- nitrogen + hydrogen ⇌ ammonia
- $N_2(g) + 2H_2(g) \rightleftharpoons 2NH_3(g)$

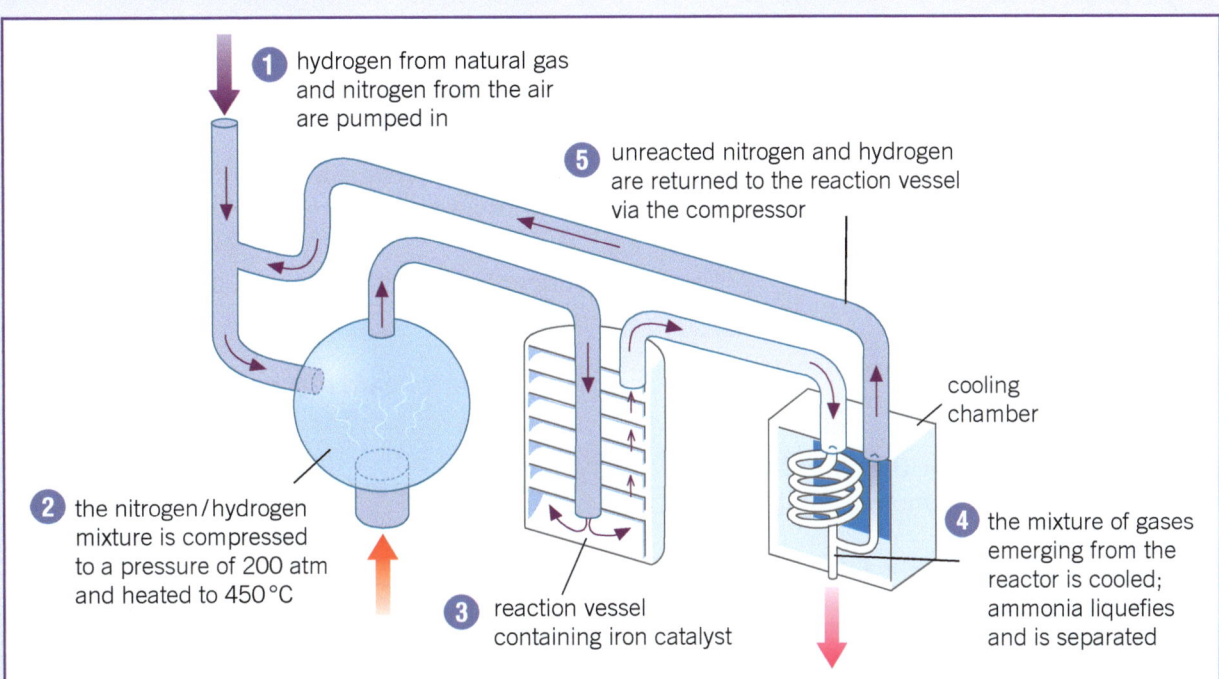

1. hydrogen from natural gas and nitrogen from the air are pumped in
2. the nitrogen/hydrogen mixture is compressed to a pressure of 200 atm and heated to 450 °C
3. reaction vessel containing iron catalyst
4. the mixture of gases emerging from the reactor is cooled; ammonia liquefies and is separated
5. unreacted nitrogen and hydrogen are returned to the reaction vessel via the compressor

Conditions

Compromise
The conditions are a *compromise* to balance yield, cost, and rate.
- an iron catalyst
- temperature ≈ 450 °C
- pressure ≈ 200 atmospheres

Catalyst
Iron is an effective catalyst for the Haber process. It does not increase the yield, but does increase the rate.

Temperature
The forward reaction is exothermic. Therefore, lowering the temperature would increase the yield of ammonia, but would also decrease the rate of reaction.

Pressure
There are fewer gas molecules on the product side, so increasing the pressure would increase the yield and the rate of reaction. However, it is very expensive to increase the pressure.

The contact process

Sulfuric acid is produced industrially using the contact process.

Stage 1: Sulfur is burned in air to produce sulfur dioxide.

$$S(s) + O_2(g) \rightarrow SO_2(g)$$

Stage 2: Sulfur dioxide reacts with more oxygen to make sulfur trioxide.

$$2SO_2(g) + O_2(g) \rightarrow 2SO_3(g)$$

This exothermic reaction is reversible and it requires a catalyst of vanadium(V) oxide, V_2O_5, a temperature of around 450 °C and atmospheric pressure.

Key Terms
Make sure you can write a definition for these key terms.

closed system   conditions   dynamic equilibrium   Haber process   reversible reaction

Retrieval

Learn the answers to the questions below then cover the answers column with a piece of paper and write down as many as you can. Check and repeat.

C10 questions | Answers

#	Question	Answer
1	What is a reversible reaction?	one where the reactants turn into products and the products turn into reactants
2	Which symbol shows a reversible reaction?	$\rightleftharpoons$
3	What is a closed system?	A reaction where no products or reactants are allowed to escape
4	What is dynamic equilibrium?	the point in a reversible reaction when the rates of the forward and reverse reactions are the same
5	What are the three reaction conditions that can be changed?	concentration, temperature, pressure
6	What is the effect of increasing the concentration of reactants on a reaction at dynamic equilibrium?	favours the forward reaction
7	What is the effect of decreasing the concentration of products on a reaction at dynamic equilibrium?	favours the forward reaction
8	What is the effect of increasing the temperature on a reversible reaction at equilibrium?	favours the endothermic direction
9	What is the effect of decreasing the temperature on a reversible reaction at equilibrium?	favours the exothermic direction
10	What is the effect of increasing the pressure on a reversible reaction at equilibrium?	favours the side with the fewest moles of gas
11	What is the effect of adding a catalyst on the position of equilibrium?	no effect, just increases the rate in both directions
12	What is the product of the Haber process?	ammonia
13	What is ammonia used for?	fertilisers
14	What is the catalyst used in the Haber process?	iron
15	What are the conditions used in the Haber process?	450 °C, 200 atm, iron catalyst
16	Write a balanced symbol equation for the Haber process.	$N_2(g) + 3H_2(g) \rightleftharpoons 2NH_3(g)$

C10

Now use the questions below to check your knowledge from previous chapters.

Previous questions / Answers

#	Question	Answer
1	What is collision theory?	that reactants need to collide in order to react
2	What is the activation energy?	the amount of energy colliding particles have to have before a reaction will take place
3	Describe the effect of increasing concentration on the rate of reaction	increases
4	Explain the effect of increasing concentration on the rate of reaction	more reactant particles lead to more frequent collisions
5	What is a mole?	a number of particles
6	What is Avogadro's number?	6.022×10^{23}
7	What formula relates moles, mass, and M_r?	moles = mass/M_r
8	What is the unit for concentration?	g/dm^3 or mol/dm^3
9	Which formula relates concentration, mass, and volume?	concentration = mass/volume
10	What is the test for copper(II) ions?	form a blue precipitate with sodium hydroxide
11	What is the test for iron(II) ions?	form a green precipitate with sodium hydroxide
12	What is the test for iron(III) ions?	form a brown precipitate with sodium hydroxide
13	What is the test for a chloride ion?	add silver nitrate and nitric acid, white precipitate
14	What is the test for a bromide ion?	add silver nitrate and nitric acid, cream precipitate
15	What is the test for an iodide ion?	add silver nitrate and nitric acid, yellow precipitate

(Put paper here)

Maths Skills

Practise your maths skills using the worked example and practice questions below.

Ratios, fractions, and percentages

In chemistry we often use ratios, fractions, and percentages to describe mixtures. These are different mathematical forms of numbers that represent the same thing.

A **ratio** compares the size of two or more quantities.

A **fraction** can express a part of a whole number, or shows one number divided by another in an equation.

A **percentage** is a number expressed as a fraction of 100.

Worked example

A stoppered flask contains a gas that is a mixture of 70 atoms of neon and 50 atoms of helium.

The **ratio** of neon atoms to helium atoms in the mixture is 70:50, which simplifies to 7:5 by dividing each side by the highest common factor (in this case, 10).

The **fraction** of atoms that are neon is:

$$\frac{70}{(70+50)} = \frac{70}{120} = \frac{7}{12}$$

The **percentage** of atoms that are helium is:

$$\left(\frac{50}{(70+50)}\right) \times 100 = 41.67\%$$

Practice

1. An equilibrium mixture contains $45\,cm^3$ of H_2 and $22.5\,cm^3$ of O_2. What is the ratio of H_2 to O_2?

2. What fraction of the total volume in this mixture is O_2?

3. A different equilibrium mixture contains $92\,cm^2$ of H_2, $154\,cm^2$ of N_2, and $23\,cm^3$ of NH_3. What is the ratio of the three different substances?

4. What is the percentage of NH_3 and N_2 combined out of the whole mixture?

Practice

Exam-style questions

01 Ammonia is formed from the reversible reaction between nitrogen and hydrogen.

$$N_2(g) + 3H_2(g) \rightleftharpoons 2NH_3(g)$$

01.1 The forward reaction transfers 92 kJ of energy to the surroundings. State how much energy is transferred by the reverse reaction. **[1 mark]**

> **Exam Tip**
> Energy is always conserved.

_____

01.2 Describe why this reaction must occur within a closed system. **[1 mark]**

_____
_____
_____

01.3 The reacting mixture is placed in apparatus that prevents the escape of reactants and products.

The pressure of the reaction mixture is then increased.

Describe what happens to the position of the equilibrium. **[1 mark]**

Tick **one** box.

it does not change ☐

it shifts to the left ☐

> **Exam Tip**
> Look at the number of moles on each side of the reaction.

it shifts to the right ☐

it shifts to the left and then to the right ☐

02 Sulfur dioxide reacts with oxygen in a reversible reaction.

$$2SO_2(g) + O_2(g) \rightleftharpoons 2SO_3(g)$$

The forward reaction is exothermic.

02.1 Draw a dot and cross diagram to show the bonding in an oxygen molecule, O_2. **[2 marks]**

02.2 Define the term exothermic. **[1 mark]**

_____
_____

C10 The extent of reactions

C10

02.3 The reversible reaction above occurs in a closed container.
Give the effect on the position of the equilibrium for each of the following condition changes. **[4 marks]**

more SO_3 is added: _____

pressure is increased: _____

temperature is increased: _____

more O_2 is added: _____

! Exam Tip

The equilibrium position shifts to counter the change; this means the reaction will go in the opposite direction to the change made.

03 Methanol is used as a fuel. It can be produced by reacting carbon monoxide with hydrogen in a reversible reaction.
$$CO(g) + 2H_2(g) \rightleftharpoons CH_3OH(g)$$
The forward reaction is exothermic.

03.1 Explain why this reaction can only reach equilibrium in a sealed container. **[3 marks]**

03.2 Identify **three** factors that will affect the position of the equilibrium reaction to produce methanol. Explain the effect that changing each of these factors has on the position of the equilibrium. **[6 marks]**

! Exam Tip

This is a 6 mark question asking for three factors; you will get one mark for each factor and one mark for explaining its effect on the equilibrium.

03.3 Calculate the maximum mass of methanol that can be made from 10.0 g of carbon monoxide. Give your answer to three significant figures. Use the Periodic Table to help you. **[6 marks]**

04 Hydrogen reacts with iodine to form hydrogen iodide.
$$H_2(g) + I_2(g) \rightleftharpoons 2HI(g)$$

04.1 State what the $\rightleftharpoons$ symbol tells you about the reaction. **[1 mark]**

04.2 The reaction reaches equilibrium in apparatus that prevents the escape of reactants and products. Describe what happens to the particles of H_2, I_2, and HI at equilibrium. **[1 mark]**

04.3 The forward reaction is endothermic. Describe the energy transfers involved in the forward reaction. **[2 marks]**

! Exam Tip

Use the correct keywords to get the marks in this answer.

05 Some students want to investigate the reversible change:
hydrated copper sulfate $\rightleftharpoons$ anhydrous copper sulfate + water

Figure 1 shows the apparatus.

Figure 1

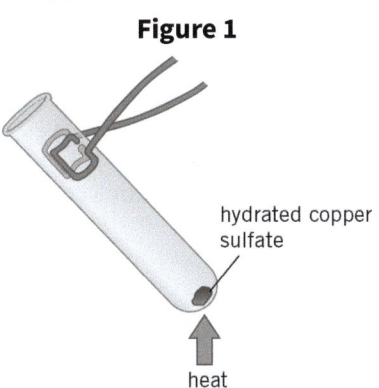

hydrated copper sulfate

heat

05.1 Explain why equilibrium cannot be reached using the apparatus in **Figure 1**. **[1 mark]**

05.2 Suggest a suitable piece of equipment for heating the hydrated copper sulfate. **[1 mark]**

C10 Practice 117

05.3 Name the substance that leaves the test tube. **[1 mark]**

05.4 Suggest how the substance that leaves the test tube could be collected. **[3 marks]**

> **! Exam Tip**
> You only need to suggest how to collect it, not test it.

06 A student sets up an equilibrium of two nitrogen oxides in a sealed gas syringe. The equilibrium is represented by the equation:

$$N_2O_4(g) \rightleftharpoons 2NO_2(g)$$
colourless   brown

At equilibrium the substances in the syringe are light brown in colour. The student recorded their observations when they moved the plunger of the gas syringe (**Table 1**).

Table 1

Action	Colour change
push in the syringe plunger	light brown to colourless
pull out the syringe plunger	light brown to dark brown

> **! Exam Tip**
> Look at the number of moles on each side of the equation.

06.1 Explain what happens to the equilibrium when the plunger is pushed in. **[4 marks]**

> **! Exam Tip**
> Remember this is a reversible reaction, so the reaction will change direction to counter the change made in the environment.

06.2 The forward reaction is endothermic. The student places the syringe in an ice–water mixture. Predict and explain the colour change observed. **[3 marks]**

06.3 The student sets another equilibrium up in a separate gas syringe:

$$H_2(g) + I_2(g) \rightleftharpoons 2HI(g)$$

Predict and explain the effect on the position of equilibrium of increasing the pressure on the equilibrium mixture. **[2 marks]**

> **! Exam Tip**
> Don't use abbreviations or short hand in your answer; in class you might be used to LHS for left-hand side but in the exam you need to write out the words in full.

07 The reaction between two substances, **X** and **Y**, is reversible:

$$X(g) \rightleftharpoons Y(g)$$

Substance **Y** is placed in a sealed container. After some time, equilibrium is established. **Figure 2** shows how the concentrations of **X** and **Y** change as equilibrium is established.

Figure 2

(graph: Concentration vs Time in min, 0–11; curves labelled substance X and substance Y)

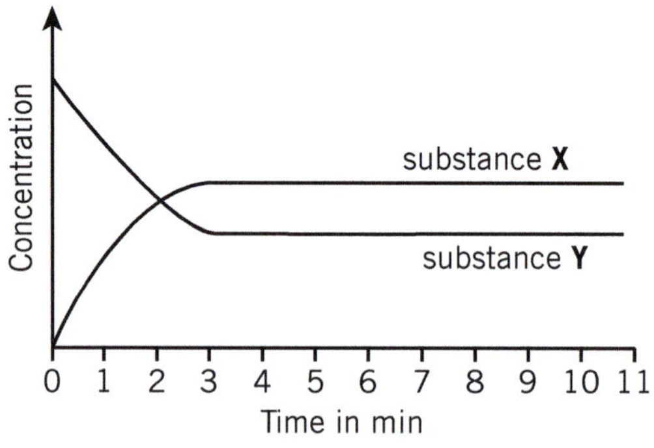

> **! Exam Tip**
> This is a different example but the same principles apply. It might seem like a trick question but it is just testing if you really understand what is going on.

07.1 Identify the time at which the equilibrium was established. Give a reason for your answer. **[2 marks]**

07.2 State how the rate of the forward and reverse reactions compares at equilibrium. **[1 mark]**

> **! Exam Tip**
> Use data from the graph to support your answer.

07.3 The system was heated for 5 minutes. At the end of the 5 minutes, it was found that there was more of substance **X** in the system than before the system was heated. Identify which reaction is exothermic. **[1 mark]**

08 A teacher had a closed system at equilibrium that contained Cl_2 (a pale green gas), ICl_3 (yellow crystals), and ICl (a brown liquid). The teacher removed Cl_2 from the system, and left the system to reach equilibrium. At the new equilibrium, the amount of ICl had increased and the amount of ICl_3 had decreased.

08.1 Write a balanced symbol equation with state symbols for the reaction. **[3 marks]**

> **Exam Tip**
>
> Use two colours to highlight the text. Pick out compounds on the left-hand side in one colour and compounds on the right-hand side in another.

08.2 The teacher placed the system into an ice bath. More yellow crystals formed. Identify whether the formation of ICl_3 is exothermic or endothermic. **[1 mark]**

08.3 Explain why placing the system in an ice bath favours this reaction. **[3 marks]**

09 A student heats a sample of ammonium chloride in a test tube. The ammonium chloride breaks down into ammonia gas and hydrogen chloride gas.

solid ammonium chloride ⇌ ammonia gas + hydrogen chloride gas

As the student heats the ammonium chloride, a white solid forms at the top of the test tube (**Figure 3**).

Figure 3

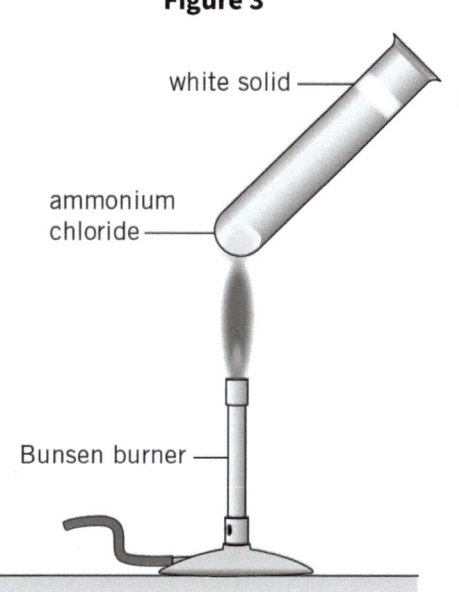

09.1 Predict what the white solid is. **[1 mark]**

09.2 Explain your prediction in **09.1**. **[3 marks]**

> **Exam Tip**
>
> Look for the clues in the main body of the question.

09.3 Another student heats hydrated blue copper(II) sulfate crystals in a test tube. Water is given off to form white anhydrous copper(II) sulfate crystals. This process is reversible. Predict what would happen if water was added to the anhydrous copper sulfate crystals. **[1 mark]**

09.4 The student holds the test tube while they add the water. Predict what they would feel as water is added. **[1 mark]**

10 Sulfuric acid is produced in a multi-step process called the contact process. One step within the contact process involves the reaction between sulfur dioxide and oxygen.

10.1 Complete and balance the symbol equation for the reaction between sulfur dioxide and oxygen. **[3 marks]**

_____(g) + O₂(_____) ⇌ _____SO₃(g)

> **Exam Tip**
> Only write in the gaps! Do not try to add in extra numbers or letters outside of the gaps.

10.2 Increasing the temperature of the reaction vessel causes the equilibrium position to shift to the left. Identify which reaction is exothermic. **[1 mark]**

10.3 Describe what this suggests about the energy transfers involved in the chemical reaction. **[1 mark]**

10.4 In industry, the reaction is carried out at 450 °C rather than room temperature. Suggest what effect this will have on the yield of sulfur trioxide. **[1 mark]**

10.5 Suggest why the reaction is carried out at 450 °C. **[1 mark]**

10.6 Vanadium pentoxide is used as a catalyst for the forward reaction. Explain how the catalyst increases the rate of the forward reaction. **[3 marks]**

10.7 Suggest and explain **one** other condition that would favour the formation of sulfur trioxide. **[3 marks]**

10.8 Suggest a reason why this condition may not be used in the industrial process. **[1 mark]**

11 The equation shows a chemical reaction:

ammonium chloride ⇌ ammonia + hydrogen chloride

11.1 Give the meaning of the symbol ⇌. **[1 mark]**

11.2 The reaction occurs from left to right if ammonium chloride is heated. State the condition required for the reverse reaction to occur. **[1 mark]**

11.3 State **one** feature of the apparatus required for an equilibrium to be established between ammonium chloride, ammonia, and hydrogen chloride. **[1 mark]**

> **Exam Tip**
> Working out the states will help with this question.

12 The Haber process is used to produce ammonia.

12.1 Give the balanced symbol equation for the reaction. **[2 marks]**

12.2 Give **one** use for the ammonia produced in the Haber process. **[1 mark]**

12.3 Give **one** source of the nitrogen in the Haber process. **[1 mark]**

> **Exam Tip**
> When writing equations, start by writing down the chemical formula of the three substances, then balance the equation.

12.4 Give the conditions used in the industrial Haber process. Explain how the conditions make the Haber process economically viable in an industrial setting. **[6 marks]**

> **Exam Tip**
> This is a popular question as it ties in lots of chemistry from other units.

13 Ammonia is an important chemical that is produced industrially in the Haber process. **Figure 4** shows how the percentage yield of ammonia changes with different reaction conditions.

Figure 4

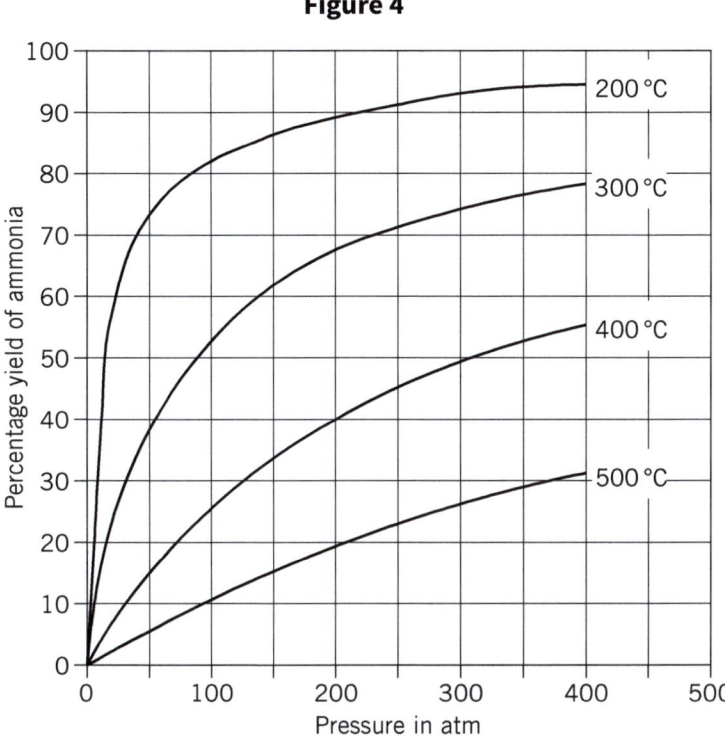

13.1 Write the balanced symbol equation with state symbols for the reaction in the Haber process. **[3 marks]**

> **Exam Tip**
> You've been asked to give the state symbols, so don't forget them! Many students do, and this is an easy mark.

13.2 Sketch a data line on **Figure 4** for the temperature that the Haber process is carried out at industrially. **[1 mark]**

13.3 Use your data line from **13.2** to identify the expected percentage yield of ammonia in the industrial Haber process. **[1 mark]**

13.4 Use **Figure 4** and your knowledge of the Haber process to explain why the industrial conditions of the Haber process are described as a compromise. **[6 marks]**

13.5 Describe the role of ammonia in food production. **[2 marks]**

14 This question is about reversible reactions.

14.1 Choose the correct words or symbols from the box to complete the sentences. Each word or symbol can be used once, more than once, or not at all. **[3 marks]**

reactants products ⇌ ⇔

In a chemical reaction, the ―――――― react together to form the ――――――.

When the equation for a chemical reaction contains the ―――――― symbol, this means that the ―――――― can also react to reform the ――――――.

14.2 Which of the following is **not** a reversible process? Choose **one** answer. **[1 mark]**

combustion of paper

ice melting

thermal decomposition of ammonium chloride

water condensing on a mirror

> **! Exam Tip**
>
> Reversible means being able to undo something. Which of these reactions cannot be undone?

14.3 Name the state that is reached when a reversible reaction is carried out within a sealed apparatus where the reactants and products cannot escape. **[1 mark]**

14.4 The forwards reaction of a reversible reaction is exothermic. State whether the backwards reaction is exothermic or endothermic. **[1 mark]**

15 When heated, blue hydrated copper sulfate crystals form white anhydrous copper sulfate.

15.1 Identify whether this change is endothermic or exothermic. **[1 mark]**

15.2 Give **two** observations you would observe if you added water to anhydrous copper sulfate crystals. **[2 marks]**

15.3 Identify the type of bonding between the copper and sulfate ions in copper sulfate. **[1 mark]**

15.4 Suggest **one** safety precaution that should be taken when heating hydrated copper sulfate crystals. **[1 mark]**

> **! Exam Tip**
>
> You are asked for two observations in **15.2**. For one observation, you need to use the information from the introduction. For the other observation, you need to use the information from **15.1**.

C10 The extent of reactions

16 A student carries out a reaction and finds that 4 J of energy is released.

16.1 Identify whether the reaction is exothermic or endothermic. **[1 mark]**

16.2 Identify **one** way in which the student can measure the energy released. **[1 mark]**

16.3 The reaction is reversible. Complete the sentence.
As the forward reaction is ——————, the reverse reaction will be ——————.

16.4 Determine how much energy the reverse reaction will take in. **[1 mark]**

17 Dinitrogen tetroxide, N_2O_4, can form nitrogen dioxide, NO_2.

17.1 Complete the balanced symbol equation for this process. **[1 mark]**
$$N_2O_4 \rightleftharpoons \text{——————} NO_2$$

17.2 The forward reaction is endothermic. Define the term endothermic. **[1 mark]**

17.3 Dinitrogen tetroxide is a colourless gas, whereas nitrogen dioxide is a brown gas. A teacher has a mixture of both gases in a sealed container. When the container is heated, the colour of the mixture starts to look darker brown. Suggest why. **[1 mark]**

17.4 For an equilibrium to be achieved, the gases must be in a sealed container. Suggest why. **[1 mark]**

> **Exam Tip**
>
> Don't worry if you haven't heard of dinitrogen tetroxide before. Unfamiliar contexts will be used in an exam to test if you can apply your knowledge to a new situation.

Knowledge

C11 Energy changes

Energy changes

During a chemical reaction, energy transfers occur. Energy can be transferred:
- to the surroundings – **exothermic**
- from the surroundings – **endothermic**

This energy transfer can cause a temperature change. Energy is always conserved in chemical reactions. This means that there is the same amount of energy in the Universe at the start of a chemical reaction as at the end of the chemical reaction.

The surroundings

When chemists say energy is transferred from or to "the surroundings" they mean "everything that isn't the reaction".

For example, imagine you have a reaction mixture in a test tube. If you measure the temperature in the test tube using a thermometer, the thermometer is then part of the surroundings.

- If the thermometer records an increase in temperature, the reaction in the test tube is exothermic.
- If the thermometer records a decrease in temperature, the reaction in the test tube is endothermic.

Reaction profiles

A **reaction profile** shows whether a reaction is exothermic or endothermic.

The **activation energy** is the minimum amount of energy that particles must have to react when they collide.

Remember, a catalyst lowers the activation energy by providing a different pathway for the reaction to occur.

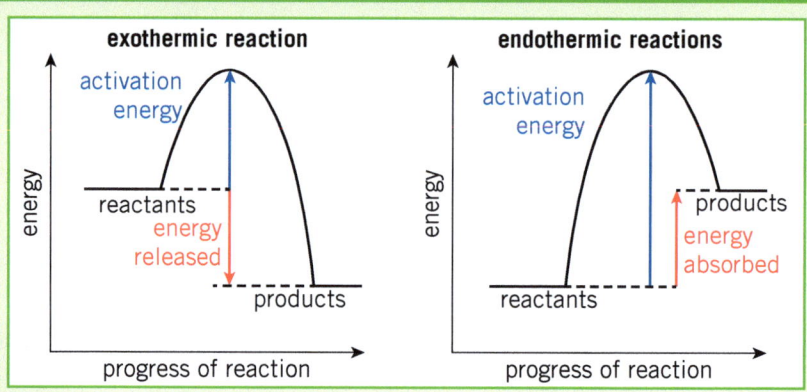

Bonds

Atoms are held together by strong chemical bonds. In a reaction, those bonds are broken and new ones are made between different atoms.
- Breaking a bond requires energy so is endothermic.
- Making a bond releases energy so is exothermic.

Breaking bonds

If a lot of energy is released when making the bonds and only a little energy is required to break them, then overall energy is released and the reaction as a whole is exothermic.

Making bonds

If a little energy is released when making the bonds and a lot is required to break them, then overall energy is taken in and the reaction as a whole is endothermic.

Bond calculations

Different bonds require different amounts of energy to be broken (their **bond energies**). To work out the overall energy change of a reaction, you need to:

1. work out how much energy is required to break all the bonds in the reactants
2. work out how much energy is released when making all the bonds in the products

overall energy transferred = energy required to break bonds − energy required to make bonds

- A positive number (+ΔH) means an endothermic reaction.
- A negative number (−ΔH) means an exothermic number.

Revision Tip

Drawing everything out helps to count the correct number of bonds.

Key Terms

Make sure you can write a definition for these key terms.

activation energy   battery   bond energy   chemical cell   combustion   endothermic
exothermic   fuel cell   neutralisation   oxidation   reaction profile   rechargeable

C11

Reaction	Energy transfer	Temperature change	Enthalpy change value (ΔH)	Example	Everyday use	Bonds
exothermic	to the surroundings	temperature of the surroundings increases	negative (−ΔH)	• **oxidation** • **combustion** • **neutralisation**	• self-heating cans • hand warmers	more energy released when making bonds than required to break bonds
endothermic	from the surroundings	temperature of the surroundings decreases	positive (+ΔH)	• **thermal decomposition** • citric acid and sodium hydrogen carbonate	• sports injury packs	less energy released when making bonds than required to break bonds

Chemical cells

In a metal displacement reaction, one metal is oxidised – it loses electrons. These electrons are transferred to another metal, which gains the electrons and is reduced.

By using a **chemical cell** to conduct this reaction, the electron's movement generates a current.

In the cell shown, the zinc atoms from the electrode lose electrons, turn into ions, and move into the solution. The electrons travel through the circuit to the copper electrode, causing the LED to light up. Once at the copper electrode, a metal ion *from the solution* will pick the electrons up and become a metal atom.

The greater the difference in reactivity between the two metals in the cell, the greater the potential difference produced.

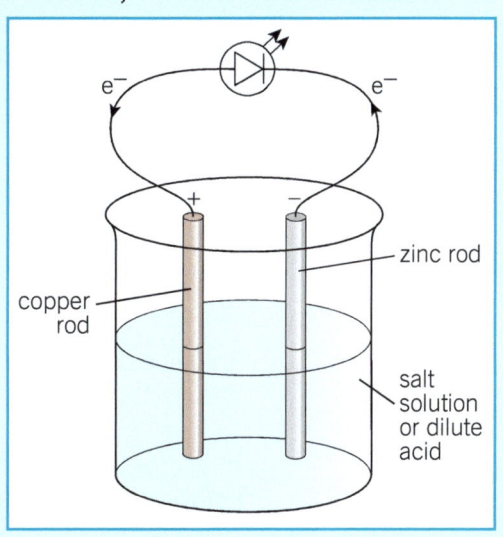

Calorimetry

We calculate the energy released from the combustion of fuels using calorimetry.

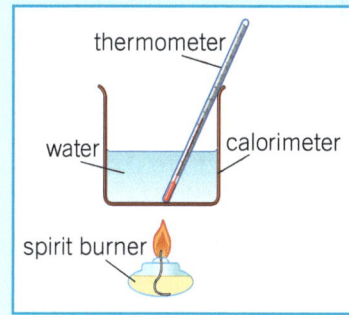

The energy released heats water in a metal can (a calorimeter). We measure how much the water temperature increases and then calculate the energy released using the equation $Q = mc\Delta T$

Hydrogen fuel cells

Fuel cells use a fuel and oxygen from the air to generate a potential difference.

Hydrogen fuel cells generate electricity from hydrogen and oxygen. The overall reaction is:

$$2H_2(g) + O_2(g) \rightarrow 2H_2O(l)$$

The hydrogen is oxidised to produce water.

Advantages
- the only waste is water
- do not need to be electrically recharged

Disadvantages
- hydrogen is highly flammable and difficult to store
- hydrogen is often produced from non-renewable resources

Specific heat capacity

When a substance is heated or cooled the temperature change depends on:
- the substance's mass
- the type of material
- how much energy is transferred to it.

Every type of material has a specific heat capacity – the amount of energy needed to raise the temperature of 1 kg of the substance by 1 °C.

The energy transferred to the substance can be calculated using the equation:

energy change (J) = mass (kg) x specific heat capacity (J/kg °C) x temperature change (°C)

$$Q = mc\Delta T$$

Retrieval

Learn the answers to the questions below then cover the answers column with a piece of paper and write as many as you can. Check and repeat.

C11 questions | Answers

#	Question	Answer
1	In terms of energy transfer, what is an exothermic reaction?	one that transfers energy to the surroundings
2	What happens to the temperature of the surroundings in an endothermic reaction?	decreases
3	Give an example of an exothermic reaction	combustion, oxidation, neutralisation
4	What is a reaction profile?	diagram showing how the energy changes in a reaction
5	What is the activation energy?	minimum amount of energy required before a collision will result in a reaction
6	Is bond breaking endothermic or exothermic?	endothermic
7	What is bond energy?	the energy required to break a bond or the energy released when a bond is formed
8	In terms of bond breaking and making, what is an exothermic reaction?	one where less energy is required to break the bonds than is released when making the bonds
9	In an endothermic reaction, will the overall energy change be positive or negative?	positive
10	What is a chemical cell?	a device that uses chemical reactions to generate electricity
11	How are chemical cells made?	by connecting two different metals (electrodes) in a solution (electrolyte)
12	What is a battery?	two or more chemical cells connected in series
13	How does the potential difference of a cell depend on the metals that the electrodes are made of?	the bigger the difference in reactivity, the greater the potential difference
14	How can some cells be recharged?	by applying an external current
15	Give an example of a non-rechargeable battery	alkaline batteries
16	What is a fuel cell?	a cell that uses a fuel and oxygen to generate electricity
17	Give an advantage of the hydrogen fuel cell.	only product is water, do not need to be electrically recharged
18	Give a disadvantage of the hydrogen fuel cell.	hydrogen is flammable, difficult to store, and is often produced from non-renewable sources

C11

Now use the questions below to check your knowledge from previous chapters.

Previous questions | Answers

#	Previous questions	Answers
1	In electrolysis, where are metals formed?	at the cathode
2	How can ionic substances be electrolysed?	by melting or dissolving them
3	In the electrolysis of aluminium oxide, why is the aluminium oxide mixed with cryolite?	to lower the melting point
4	In terms of pH, what is an acid?	a solution with a pH of less than 7
5	What is a neutral solution?	a solution with a pH of 7
6	How is the amount of H^+ in a solution related to its pH?	more H^+, lower pH
7	What is an ion?	an atom that has lost or gained electrons
8	How are metals that are more reactive than carbon extracted?	electrolysis
9	Why is gold found as a native metal in the ground?	it is unreactive
10	What do we call the process of removing oxygen from a compound?	reduction
11	What force holds metal ions and the delocalised electrons together?	the electrostatic force of attraction
12	What is the test and result for an iodide ion?	add silver nitrate and nitric acid, yellow precipitate
13	What is the test and result for a sulfate ion?	add hydrochloric acid and barium chloride, white precipitate

Maths Skills

Practise your maths skills using the worked example and practice questions below.

Orders of magnitude

An order of magnitude is when you look at the difference in values with reference to powers of ten.

For example, 200 is larger than 20 by 180, but 200 is *one order of magnitude larger* because $20 \times 10 = 200$.

Similarly, 7000 is three orders of magnitude larger than 7.

Worked example

A piece of marble has a length of 60 cm. A smaller piece of marble has a length of 0.6 cm. By how many orders of magnitude do they differ in size?

Step 1: Divide the bigger number by the smaller one:

$$\frac{60}{0.6} = 100$$

Step 2: Count the zeros – 100 has two zeros.

Step 3: The number of zeros = orders of magnitude, so the marbles differ in size by two orders of magnitude.

Practice

1. A square has a length of 40 cm. A larger square has a length of 40 000 cm. By how many orders of magnitude is the larger square longer than the smaller one?

2. A solution contains 20 000 hydrogen ions. After a reaction, it contains only 20 hydrogen ions. By how many orders of magnitude does the number of hydrogen ions differ before and after the reaction?

Practice

Exam-style questions

01 This question is about endothermic and exothermic reactions.

01.1 Draw **one** line from each reaction to show whether it is endothermic or exothermic. **[4 marks]**

Reaction	Endothermic or exothermic
thermal decomposition
citric acid with sodium hydrogencarbonate	endothermic
neutralisation	exothermic
combustion

> **Exam Tip**
>
> Read the instructions carefully; you may only see the word **one** in bold and think you need to draw one line but read the whole thing and you'll see that you need **four** lines, one from each box on the left-hand side.

01.2 Identify which statement is true for an exothermic reaction. **[1 mark]**

Tick **one** box.

- It transfers energy to the surroundings. ☐
- It transfers energy from the surroundings. ☐
- The energy of the products is higher than the energy of the reactants. ☐
- The temperature of the surroundings decreases. ☐

01.3 Iron oxide reacts with aluminium to produce aluminium oxide and iron. The reaction occurs at a high enough temperature that the iron produced is molten.

Identify whether the reaction is exothermic or endothermic.

[1 mark]

_____

02 Hydrogen reacts with chlorine to form hydrochloric acid. The reaction is exothermic.

Figure 1 shows the reaction profile for the reaction.

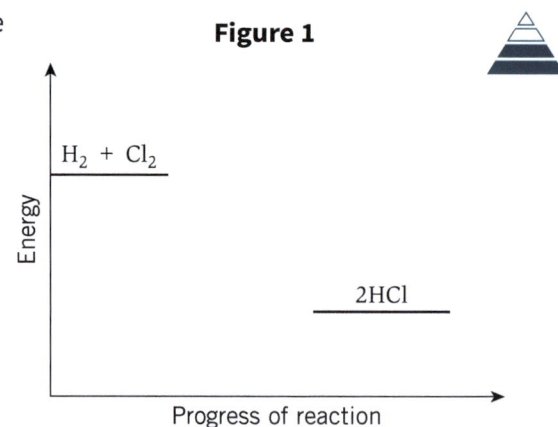

Figure 1

128 C11 Energy changes

C11

02.1 Complete the reaction profile to show how the energy changes as the reaction proceeds. Draw an arrow to show the overall energy change of the reaction.

The arrow and line do **not** need to be to scale. **[2 marks]**

02.2 Explain how the reaction profile in **Figure 1** shows that the reaction is exothermic. **[2 marks]**

_____

_____

_____

> **! Exam Tip**
> You need to refer to the position of the products and reactants in your answer.

02.3 The displayed formulae for the reaction are:

H—H + Cl—Cl → 2 H—Cl

Table 1 shows some bond enthalpies.

Table 1

	H—H	Cl—Cl	H—Cl
Energy in kJ/mol	436	242	431

Calculate the overall energy change of the reaction. **[3 marks]**

> **! Exam Tip**
> Don't forget there are 2 H—Cl bonds, even though only one has been drawn.

bond energy _____ kJ/mol

03 Some students dissolved four substances in water. This is the method used:

1. Transfer 100 cm³ of water to a beaker.
2. Measure the temperature of the water.
3. Add 1 spatula measure of the substance, in powder form, to the water.
4. Measure the new temperature of the water.

Table 2 shows their results.

Table 2

Substance	Temperature of water at start in °C	Temperature of solution immediately after dissolving in °C
A	20	25
B	21	17
C	21	31
D	22	6

> **! Exam Tip**
> You may have done this practical by mixing acid and alkali; try to remember what you put the acid into. Even though you use a beaker, this is not where the reaction occurs.

03.1 Suggest the apparatus that could be used instead of a beaker to improve step **1**. Give a reason for your suggestion. **[2 marks]**

C11 Practice 129

03.2 Suggest what the students should do between steps **3** and **4**. Give a reason for your suggestion. **[2 marks]**

> **Exam Tip**
> To work this it out it is helpful to add an extra column on the right-hand side of the table to show change in temperature.

03.3 Give the letter of the substance that dissolves most exothermically. **[1 mark]**

03.4 Predict how the temperature changes would alter if the students repeated the experiment using 200 cm³ of water in step **1**. **[1 mark]**

04 A student was investigating the burning of fuels. They placed a spirit burner under 500 cm³ of water. They measured the temperature and then lit the burner. They heated the water for 3 minutes and then recorded the maximum temperature. Their results are shown in **Table 3**.

Table 3

Temperature of 500 cm³ of water before heating in °C	22.2
Temperature of 500 cm³ of water after heating in °C	36.7

04.1 Calculate the temperature change of the water. **[1 mark]**

04.2 Calculate the amount of energy absorbed by the water. Give your answer to three significant figures. The specific heat capacity of water is 4.18 J/kg °C. **[2 marks]**

04.3 In 3 minutes, 0.29 g of fuel was burnt. Calculate the energy release per gram of fuel. **[1 mark]**

05 **Figure 2** shows the displayed formulae for the combustion reaction of methane.

Figure 2

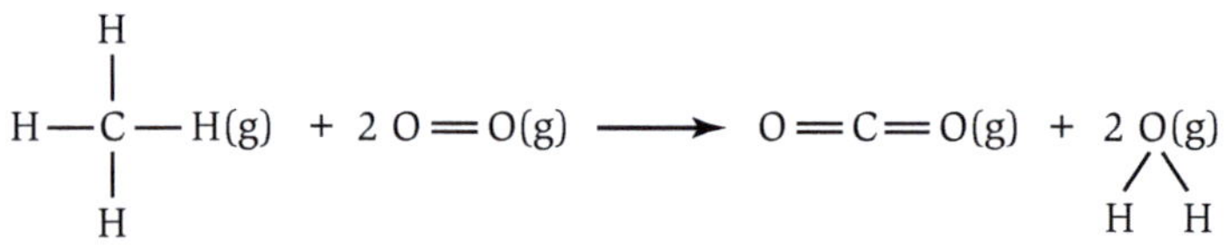

$$H-\underset{\underset{H}{|}}{\overset{\overset{H}{|}}{C}}-H(g) + 2\,O=O(g) \longrightarrow O=C=O(g) + 2\,O\underset{H\ H}{\diagup\diagdown}(g)$$

05.1 **Table 4** shows some bond enthalpy values.

Table 4

	C—H	O=O	C=O	O—H
Energy in kJ/mol	412	496	743	463

Figure 3 is the reaction profile for the reaction shown in **Figure 2**. It is **not** drawn to scale.

Calculate the energy change for each arrow **A**, **B**, and **C** shown in **Figure 3**. **[3 marks]**

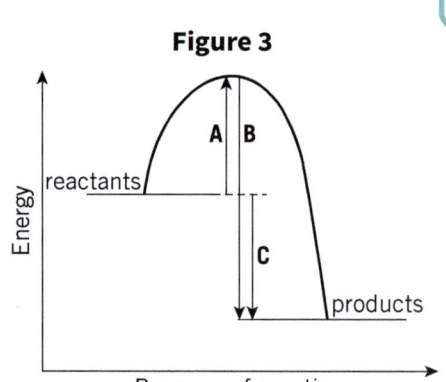

Figure 3

> **Exam Tip**
> Draw out all of the compounds in full; this will help you count all the bonds. It's easy to miss that there are FOUR O—H bonds in this equation.

C11 Energy changes

05.2 Name the energy change represented by arrow **C**. [1 mark]

05.3 Define the energy change represented by arrow **A**. [1 mark]

06 A student wanted to compare the temperature changes when different metals reacted with hydrochloric acid. The student set up the apparatus shown in **Figure 4**.

Figure 4

06.1 Name the dependent variable in the investigation. [1 mark]

06.2 Identify **two** control variables in the investigation. [2 marks]

> **Exam Tip**
> Control variables are all the parts of the experiment that we need to keep the same to make sure it's a fair test.

06.3 The student's results are shown in **Table 5**.

Table 5

Metal	Temperature of acid at start in °C	Temperature of mixture immediately after reaction in °C	Temperature change in °C
magnesium	19.0	36.7	
zinc	19.5	25.6	
copper	20.4	20.4	0.0

Suggest a reason for the result for copper. [1 mark]

06.4 Complete the missing values in **Table 5**. [1 mark]

06.5 State which metal reacts most exothermically with hydrochloric acid. [1 mark]

07 **Figure 5** shows the reaction profiles of four reactions: **A**, **B**, **C**, and **D**. The reactions profiles are drawn to scale.

Figure 5

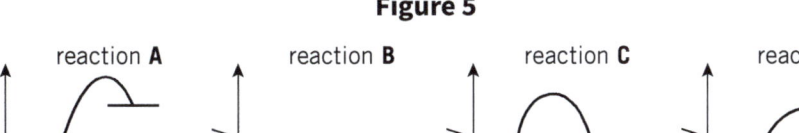

> **Exam Tip**
> To help with this question, draw arrows to show the activation energy, from where the reaction starts to its highest point.

07.1 Give the letters of **two** reaction profiles that could show combustion reactions. [1 mark]

07.2 Give the letter of the **one** reaction with the smallest activation energy. [1 mark]

07.3 Give the letter of the **one** reaction that is most exothermic. [1 mark]

08 Some students wanted to investigate the energy changes of the reactions of hydrochloric acid with three metal carbonates.

08.1 Write a balanced equation for the reaction of copper carbonate with hydrochloric acid. Include state symbols. **[3 marks]**

08.2 Describe a method that the students could use to carry out the investigation. In your answer, name the apparatus required and identify the independent, dependent, and control variables. **[6 marks]**

08.3 State how the investigation results show which metal carbonate reacts most exothermically with acid. **[1 mark]**

> **! Exam Tip**
> There are several steps in writing this equation:
> Step 1: Recall the formula for hydrochloric acid.
> Step 2: Recall the formula for a carbonate ion.
> Step 3: Determine the formula for copper carbonate.
> Step 4: Recall the general products in the equation for an acid + metal carbonate salt.
> Step 5: Determine the formula for the salt produced.
> Step 6: Balance the equation.

09 A student makes an electrical cell by connecting two different metals with a salt solution.

09.1 Identify which of the metals below would produce the largest potential difference when connected with copper.
Explain your answer. **[2 marks]**

iron   lead   tin

09.2 Suggest why potassium is not a suitable metal to use. **[1 mark]**

09.3 The student uses zinc as metal **A**. Write an ionic equation for the oxidation reaction that will happen in this chemical cell. **[3 marks]**

10 This question is about two fuels. **Table 6** gives the energy released on burning 1 mole of each of the fuels.

Table 6

Fuel name	Chemical formula of fuel	Energy transferred to the surroundings per mole of fuel burnt in kJ/mol
methane	CH_4	890
nonane	C_9H_{20}	6125

10.1 The equation for the combustion of methane is:
$$CH_4 + 2O_2 \rightarrow CO_2 + 2H_2O$$
Identify which substance is oxidised in the reaction. **[1 mark]**

10.2 The combustion products of nonane are the same as those for methane. Write a balanced chemical equation for the combustion of nonane. **[3 marks]**

10.3 Compare the environmental impacts of burning the two fuels in terms of the energy transferred per gram of fuel burnt and the energy transferred per gram of carbon dioxide produced. Relative atomic masses A_r: C = 12; H = 1; O = 16 **[6 marks]**

> **! Exam Tip**
> The numbers in the equation are getting large; always balance oxygen last as there are so many of those. If you try to start with oxygen you risk getting confused. Start with carbon, then hydrogen, and the oxygen will then be easier to sort out.

C11 Energy changes

C11

11 Some students investigated the equilibrium reaction between two solutions containing cobalt ions:

pink cobalt ion solution + chloride ions ⇌ blue cobalt ion solution + water

Figure 6 shows the apparatus.

Figure 6

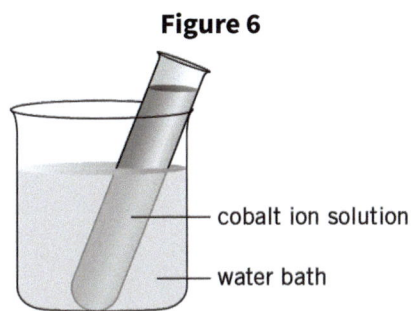

11.1 Suggest an improvement to the apparatus to make sure that all the cobalt ion solution is at the same temperature. **[1 mark]**

11.2 The students carry out some tests. **Table 7** shows their results.

Table 7

Test number	Action	Initial colour change
1	heat the water bath	from pink to blue
2	add ice to the water bath	from blue to pink
3	add chloride ions to the solution	
4	add water to the solution	

Write a conclusion based on the data in rows **1** and **2** in **Table 7**. **[2 marks]**

11.3 Predict the colour change that would be observed in row **3** of **Table 7**. **[1 mark]**

11.4 Explain the prediction you made to answer **11.3**. **[3 marks]**

> **! Exam Tip**
> Remember this is a reversible reaction, so the reaction will change direction to counter the change made in the environment.

12 The reaction between hydrochloric acid and sodium hydroxide is exothermic.

12.1 Identify what type of reaction this is. Choose **one** answer. **[1 mark]**

decomposition   electrolysis   neutralisation   reduction

12.2 A student makes the following conclusion about the reaction.

"As the reaction happened, the test tube got hotter. This shows that more energy was produced."

Is the student correct? Explain your answer. **[3 marks]**

12.3 Describe the energy transfers involved in the reaction between hydrochloric acid and sodium hydroxide in terms of bonds, and how this makes the reaction exothermic. **[3 marks]**

12.4 Give **one** other example of an exothermic reaction. **[1 mark]**

13 **Figure 7** shows a reaction profile.

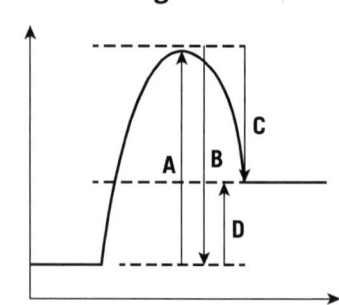

Figure 7

13.1 Write the label that should be on the *y*-axis. **[1 mark]**

13.2 Give the letter of the arrow that shows the activation energy of the reaction. **[1 mark]**

13.3 Explain whether the reaction is exothermic or endothermic. **[2 marks]**

13.4 Give **one** example of the type of reaction from **13.3**. **[1 mark]**

14 This question is about endothermic and exothermic reactions.

14.1 Tick **one** cell in **Table 8** to identify whether each reaction is endothermic or exothermic. **[4 marks]**

Table 8

Reaction	Endothermic	Exothermic
citric acid and sodium hydrogencarbonate		
combustion		
neutralisation		
thermal decomposition		

14.2 Which statement is true for an exothermic reaction? Choose **one** answer. **[1 mark]**

It transfers energy to the surroundings.

It transfers energy from the surroundings.

The energy of the products is higher than the energy of the reactants.

The temperature of the surroundings decreases.

> **Exam Tip**
> It can help if you replace the word energy with the word heat.

14.3 Iron oxide reacts with aluminium to produce aluminium oxide and iron. The reaction occurs at a high enough temperature that the iron produced is molten. Identify whether the reaction is exothermic or endothermic. **[1 mark]**

15 This question is about the reaction between iron and oxygen.

15.1 Complete the word equation for the reaction. Choose the correct answer from the box. **[1 mark]**

| iron oxygen | iron oxide | oxygen ironide | oxygen iron |

iron + oxygen → _____

15.2 Identify what type of reaction occurs between iron and oxygen.

134 C11 Energy changes

Choose **one** answer. Give a reason for your answer. **[2 marks]**

combustion   oxidation
electrolysis   neutralisation

15.3 Complete the following sentences. **[2 marks]**

A chemical reaction can only occur when the reaction particles _____ with sufficient energy.

The minimum amount of energy the particles must have is called the _____.

15.4 The temperature of the surroundings increases during the reaction between iron and oxygen. Identify whether this reaction is exothermic or endothermic. **[1 mark]**

15.5 Complete the reaction profile in **Figure 8** for the reaction between iron and oxygen. **[1 mark]**

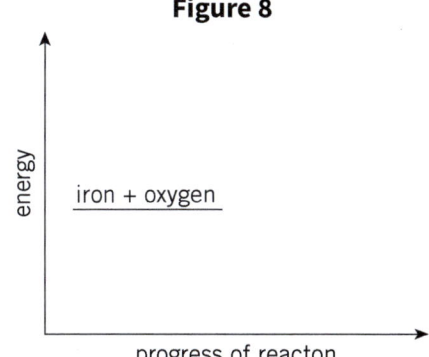

Figure 8

16 Two students mixed solutions of sulfuric acid and potassium hydroxide together and measured the resulting temperature change. Student **A** used a thermometer and student **B** used a data logger.

16.1 Suggest whether the thermometer or data logger gave the most accurate set of data. **[1 mark]**

16.2 Student **B** recorded the data shown in **Table 9**.

Table 9

	First test	Second test	Third test	Mean temperature change
Teperature change in °C	32.6	32.9	32.5	

Calculate the mean temperature change. Give your answer to one decimal place. **[1 mark]**

> **! Exam Tip**
> The answer involves a recurring digit after the decimal point. Make sure you round this number correctly.

16.3 Student **A** says:

"We need to record the temperature as soon as we mix the solutions."

Student **B** thinks this is wrong and says:

"We need to wait until the temperature has reached its highest value before recording it."

Determine which student is correct. Give a reason for your answer. **[2 marks]**

16.4 The higher the temperature of a particle, the more energy the particle has. A third student carries out the experiment between sulfuric acid and potassium hydroxide at 5 °C. The student found that there was only a very small increase in the temperature. Suggest why the third student found this result.

> **! Exam Tip**
> To answer this question you need to use ideas about rates of reaction.

Knowledge

C12 Carbon compounds as fuels

Crude oil

Crude oil is incredibly important to our society and economy.

Raw crude oil is a thick black liquid made of a large number of different compounds mixed together. Most of the compounds are **hydrocarbons** of various sizes. Hydrocarbons are molecules made of carbon and hydrogen only.

Combustion

Hydrocarbons are used as **fuels**. This is because when they react with oxygen they release a lot of energy. This reaction is called **combustion**. Complete combustion is a type of combustion where the only products are carbon dioxide and water.

Properties

Whether or not a particular hydrocarbon is useful as a fuel depends on its properties:

- **flammability** – how easily it burns
- **boiling point** – the temperature at which it boils
- **viscosity** – how thick it is

Its properties in turn depend on the length of the molecule.

Chain length	Flammability	Boiling point	Viscosity
long	low	high	high (very thick)
short	high	low	low (very runny)

Alkanes

One family of hydrocarbon molecules are called **alkanes**. Alkane molecules only have single bonds in them. The first four alkanes are:

The different alkanes have different numbers of carbon atoms and hydrogen atoms. You can always work out the molecular formula of an alkane by using C_nH_{2n+2}.

methane, ethane, propane

Revision Tip

You can check if you've drawn compounds correctly since carbon always forms four bonds and hydrogen always forms one bond.

Key Terms

Make sure you can write a definition for these key terms.

alkanes   boiling point   climate change   combustion   crude oil   feedstock
flammability   fractional distillation   fuel   global warming   hydrocarbon   viscosity

C12

Pollutants released from the combustion of fuels

Pollutant	Origin	Effect
carbon dioxide	complete combustion of fuels	**global warming**, which is contributing to **climate change**
carbon monoxide	incomplete combustion of fuels	colourless and odourless toxic gas
particulates (soot and unburnt hydrocarbons)	incomplete combustion of fuels especially in diesel engines	**global dimming**, respiratory problems, potential to cause cancer
sulfur dioxide	sulfur impurities in the fuel reacting with oxygen from the air	acid rain and respiratory problems
oxides of nitrogen	nitrogen from the air being heated near an engine and reacting with oxygen	acid rain and respiratory problems

Fractional distillation

The different hydrocarbons in crude oil are separated into fractions based on their boiling points in a process called **fractional distillation**. All the molecules in a fraction have a similar number of carbon atoms, and so a similar boiling point.

The process takes place in a fractionating column, which is hot at the bottom and cooler at the top.

The process works like this:

1. Crude oil is vapourised (turned into a gas by heating).
2. The hydrocarbon gases enter the column.
3. The hydrocarbon gases rise up the column.
4. As hydrocarbon gases rise up the column they cool down.
5. When the different hydrocarbons reach their boiling point in the column they condense.
6. The hydrocarbon fraction is collected.

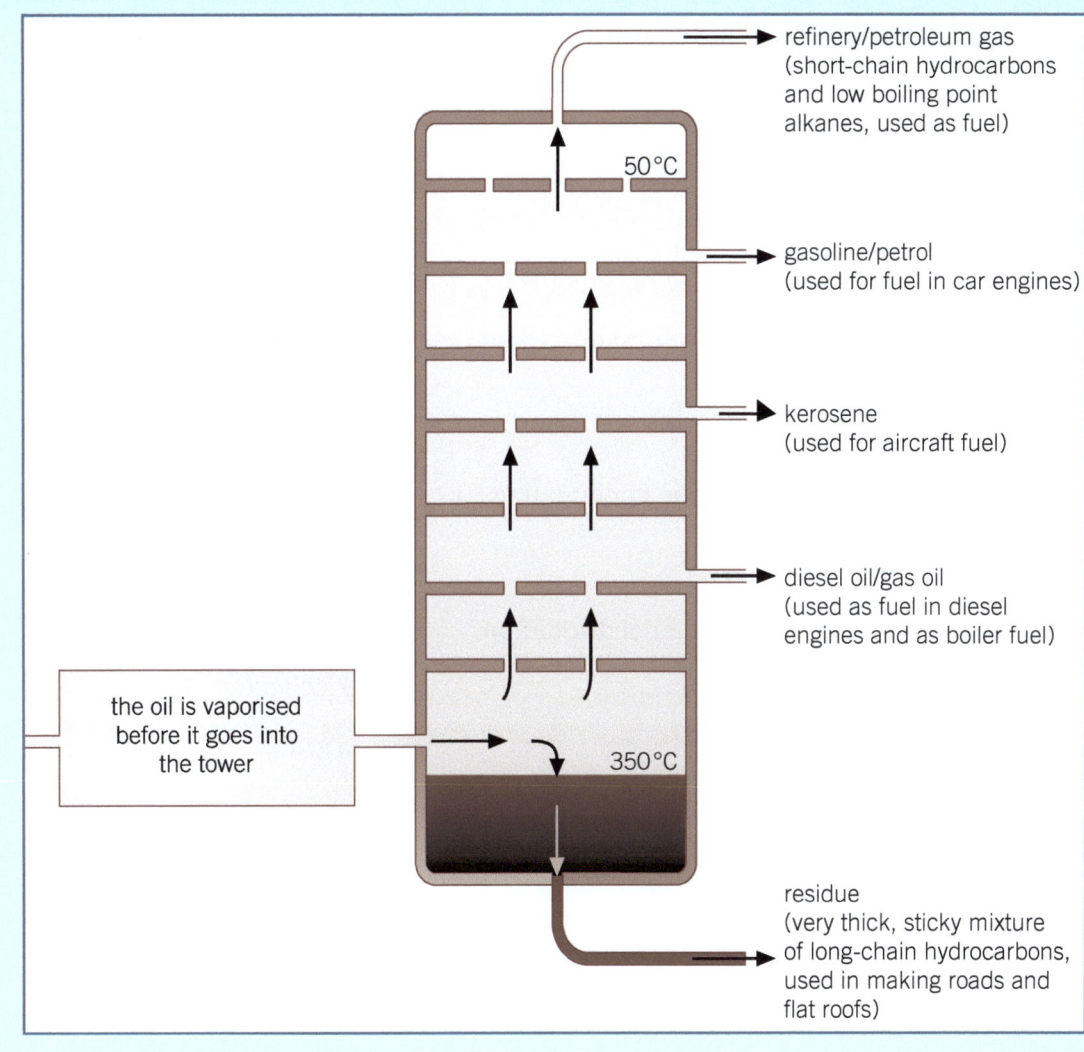

C12 Knowledge 137

Retrieval

Learn the answers to the questions below then cover the answers column with a piece of paper and write down as many as you can. Check and repeat.

C12 questions — Answers

#	Question	Answer
1	What is a hydrocarbon?	a compound containing carbon and hydrogen only
2	In terms of the substances that make it up, what is crude oil?	a mixture of different compounds, mostly hydrocarbons
3	What are the alkanes?	a family of hydrocarbons that only have single bonds
4	Name the first three alkanes.	methane, ethane, propane
5	What is the general formula for the alkanes?	C_nH_{2n+2}
6	What is a combustion reaction?	a reaction between a fuel and oxygen
7	What are the products of the complete combustion of a hydrocarbon?	carbon dioxide and water
8	What properties does an alkane's use as a fuel depend on?	boiling point, flammability, viscosity
9	What is meant by chain length?	how many carbon atoms are in an alkane
10	How does boiling point depend on chain length?	longer the chain, higher the boiling point
11	How does viscosity depend on chain length?	longer the chain, higher the viscosity
12	How does flammability depend on chain length?	longer the chain, lower the flammability
13	How can the different alkanes in crude oil be separated?	fractional distillation
14	How does the temperature in the fractionating column change?	hot at the bottom, cool at the top
15	Outline the major steps in fractional distillation.	• crude oil is vapourised • vapours rise up the column and cool down • vapours condense at their boiling points • collected as liquids
16	What is a fraction?	a group of hydrocarbons with similar chain lengths

C12

Now use the questions below to check your knowledge from previous chapters.

Previous questions / Answers

#	Question	Answer
1	Where are metals and non-metals on the Periodic Table?	metals to the left, non-metals to the right
2	What are the products of reactions of the alkali metals with oxygen, chlorine, and water?	metal oxide with oxygen, metal chloride with chlorine, metal hydroxide + hydrogen with water
3	What is the ionic equation for a reaction between an acid and an alkali?	$H^+(aq) + OH^-(aq) \rightarrow H_2O(l)$
4	How can you obtain a solid salt from a solution?	crystallisation
5	What is a salt?	compound formed from a reaction of a metal or metal containing compound with an acid
6	Explain the effect of increasing the pressure on the rate of reaction.	less space between particles means more frequent collisions so the rate increases
7	Explain the effect of increasing the surface area on the rate of reaction.	more reactant particles are exposed and able to collide, leading to more frequent collisions and the rate increases
8	Describe the effect of increasing the temperature on the rate of reaction.	increases
9	What is the empirical formula?	the simplest ration of the atoms in a compound
10	What is a titration?	a method used to calculate the concentration of an unknown solution
11	What is the yield of a reaction?	the mass of product you obtained from the reaction
12	What is the end point?	the point at which the reaction is just complete (and no substance is in excess)

Maths Skills

Practise your maths skills using the worked example and practice questions below.

2D and 3D models

Scientists often use models to describe what things look like and how they act.

These models can be 2D or 3D but they are always just approximations – they are there to help you understand but have strengths and weaknesses.

Worked example

The model shows how the layers in a metal alloy are disturbed. What are the strengths and weaknesses of this model?

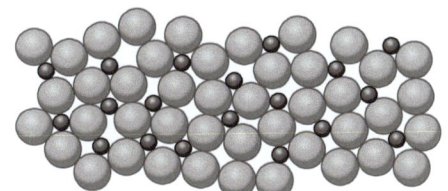

The model is in two dimensions, which helps you to see how the layers are disturbed by atoms of different sizes.

However, the metal is normally three dimensional, which this model does not show, so it is not an accurate representation of the metal's structure.

Practice

Compare and contrast the two models below showing the structure of methane.

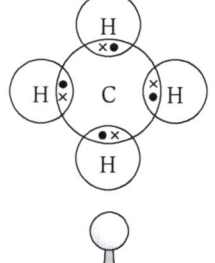

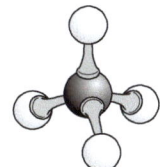

Practice

Exam-style questions

01 This question is about alkanes.

01.1 Draw **one** line from each displayed formula to the name of the alkane. **[2 marks]**

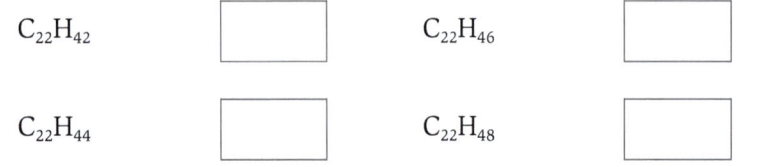

01.2 Identify the correct formula of the alkane with 22 carbon atoms. **[1 mark]**

Tick **one** box.

$C_{22}H_{42}$ ☐ $C_{22}H_{46}$ ☐

$C_{22}H_{44}$ ☐ $C_{22}H_{48}$ ☐

> **Exam Tip**
>
> Use the general formula for alkanes to figure out the number of hydrogen atoms that will be in the compound.

01.3 Decane is an alkane with 10 carbon atoms.

How do the properties of decane compare with the properties of ethane? **[1 mark]**

Tick **one** box.

Decane has higher flammability, lower boiling point, and higher viscosity. ☐

Decane has higher flammability, higher boiling point, and lower viscosity. ☐

Decane has lower flammability, higher boiling point, and higher viscosity. ☐

Decane has lower flammability, lower boiling point, and lower viscosity. ☐

> **Exam Tip**
>
> There are three possible differences within each answer; go over the properties one at a time (flammability, boiling point, and then viscosity), comparing them to ethane. This should leave you with the correct answer at the end.

140   C12 Carbon compounds as fuels

C12

02 Table 1 shows the boiling points of some alkanes.

Table 1

Name of alkane	Number of carbon atoms	Boiling point in °C
pentane	5	36
hexane	6	69
heptane	7	98
octane	8	126
nonane	9	
decane	10	174
undecane	11	196
dodecane	12	216

02.1 Plot the data from **Table 1** on **Figure 1**.
Draw a line of best fit. **[3 marks]**

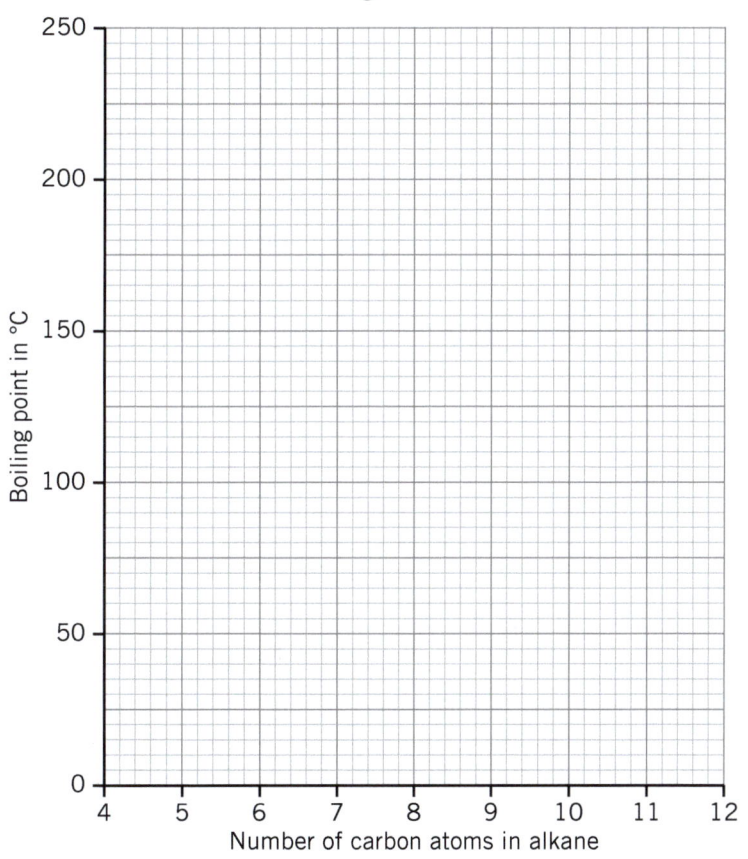

Figure 1

> **Exam Tip**
> Always plot your points with crosses – this shows the examiner exactly which point you are using.

02.2 Use your graph from **Figure 1** to predict the boiling point of nonane. **[1 mark]**

02.3 Write a chemical equation for the complete combustion of nonane. You do not need to include state symbols. **[4 marks]**

> **Exam Tip**
> The products for complete combustion are always the same two. The only things that change with large compounds is the number of moles.

C12 Practice 141

03 Many useful substances are obtained from crude oil.

03.1 Define the term fraction of crude oil. **[1 mark]**

03.2 **Figure 2** shows a fractionating column.

Explain how kerosene is separated from the other fractions in crude oil by fractional distillation. **[3 marks]**

03.3 The hydrocarbons in diesel are bigger than the hydrocarbons in petrol. Compare the physical and chemical properties of diesel and petrol. **[6 marks]**

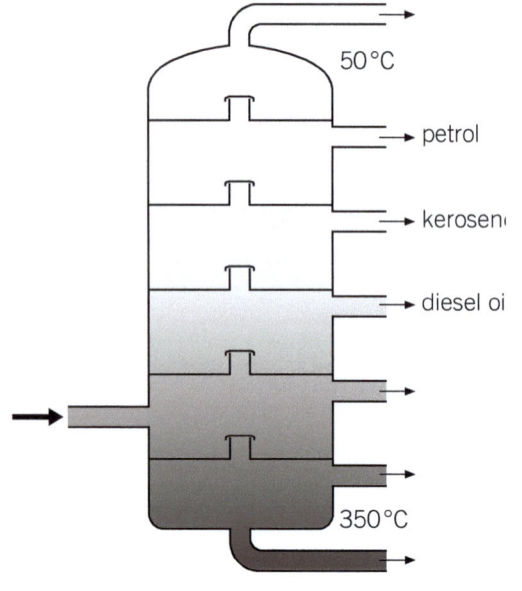

Figure 2

> **Exam Tip**
>
> Mark on the diagram where it is hottest and where it is coldest; this will point you towards the way they are separated in question **03.2** and the differences in properties in **03.3**.

04 This question is about crude oil.

04.1 Name the type of substances that crude oil is made up of. **[1 mark]**

04.2 Describe the process by which crude oil is separated into fractions. **[6 marks]**

05 **Table 2** gives the formulae and boiling points of three alkanes. The alkanes have the same numbers of carbon and hydrogen atoms, but the atoms are joined together differently. Compound **X** is a straight-chain alkane. Compounds **Y** and **Z** are branched-chain alkanes.

Table 2

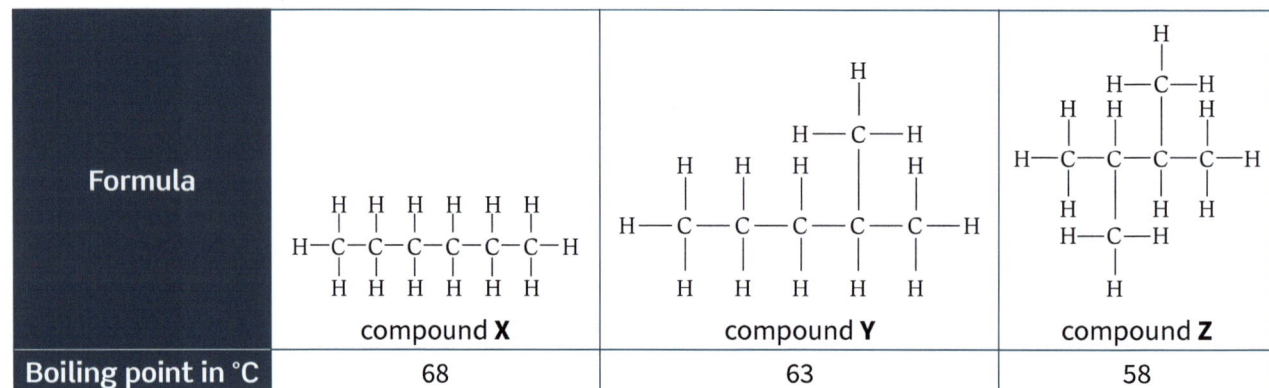

Formula	compound **X**	compound **Y**	compound **Z**
Boiling point in °C	68	63	58

05.1 Name the main source of alkanes. **[1 mark]**

05.2 Calculate the relative formula mass of the alkanes in **Table 2**. Relative atomic masses A_r: C = 12; H = 1 **[1 mark]**

05.3 Explain the relationship between boiling points and the number of branches in a molecule. **[6 marks]**

> **Exam Tip**
>
> Look at the number of branches the compounds on the previous page have and link the increase in branches to a change in boiling point.

C12

06 Some alkanes are used as fuels.

06.1 Write a balanced equation for the combustion of pentane, C_5H_{12}, to make carbon dioxide and water only. **[3 marks]**

06.2 In certain conditions, pentane undergoes incomplete combustion:

$$2C_5H_{12} + 11O_2 \rightarrow 10CO + 12H_2O$$

Deduce the conditions in which propane undergoes incomplete combustion. Use your answer to **06.1** and the equation above. Justify your answer. **[2 marks]**

06.3 Propane is another fuel. **Figure 3** shows the displayed formula equation for its complete combustion.

Figure 3

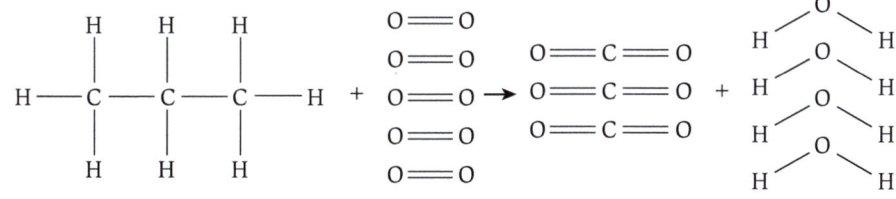

> **Exam Tip**
>
> Cross off each bond in the diagram as you count it — this should help you count everything only once and not miss bonds out.

Table 3 gives some bond enthalpy values.

Table 3

	C—C	C—H	O=O	C=O	O—H
Energy in kJ/mol	348	412	496	743	463

Calculate the energy change for the complete combustion of 1 mole of propane. **[5 marks]**

07 Crude oil is a mixture of many different substances. The substances are separated using fractional distillation.

07.1 Describe what crude oil was formed from. **[1 mark]**

07.2 Name the **two** processes that occur during fractional distillation. **[2 marks]**

08 Heptane, C_7H_{16}, is an alkane.

08.1 Name and describe the bonding between the atoms in a heptane molecule. **[2 marks]**

08.2 Name the type of force that is overcome when heptane boils. **[1 mark]**

> **Exam Tip**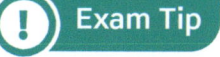
>
> The spelling of this force is important to get the marks here.

08.3 Calculate the mass of carbon dioxide produced when 85.0 g of heptane undergoes complete combustion. Give your answer to two significant figures. **[7 marks]**

C12 Practice 143

09 Petrol is used to fuel cars. **Table 4** shows the different substances that are mixed in a sample of petrol.

Table 4

Substance	Mass of substance in 200 g of petrol in g
alkanes	110
other hydrocarbons	70
ethanol	20

09.1 Use data from **Table 4** to calculate the percentage by mass of ethanol in petrol. **[2 marks]**

> **Exam Tip**
> Remember to use the general formula for alkanes.

09.2 The molecules of one alkane in petrol have seven carbon atoms. Give the formula of this alkane. **[1 mark]**

09.3 Ethanol in petrol is made from plants. The alkanes in petrol are obtained from crude oil. Suggest **one** advantage of including ethanol in petrol. **[1 mark]**

10 This question is about air pollutants.

10.1 Which pollutant is formed in car engines from the reaction between two gases that occur naturally in the atmosphere?
Choose **one** answer. **[1 mark]**

carbon dioxide   oxides of nitrogen

carbon particles   sulfur dioxide

> **Exam Tip**
> Think about which gases are found at the highest levels in the atmosphere.

10.2 Carbon monoxide is also produced in car engines.
Name the process that produces carbon monoxide. **[1 mark]**

10.3 Balance the symbol equation for the incomplete combustion of a fuel. **[1 mark]**

___ $C_4H_{10}(g)$ + ___ $O_2(g)$ → ___ $CO(g)$ + ___ $H_2O(l)$

> **Exam Tip**
> Start with the carbons, then the hydrogens and leave the oxygens until last!

10.4 Draw **one** line from each pollutant to an effect of the pollutant. **[3 marks]**

Pollutant	Effect
	poisoning of humans
oxides of nitrogen	
	global dimming
carbon monoxide	
	global climate change
particulates	
	breathing problems

> **Exam Tip**
> Not all of the boxes on the right-hand side will be used.

144 C12 Carbon compounds as fuels

10.5 Which gas causes acid rain? Choose **one** answer. **[1 mark]**

carbon dioxide   sulfur dioxide

carbon monoxide   unburnt hydrocarbons

11 Table 5 shows the boiling points of a series of hydrocarbons.

Table 5

Formula of hydrocarbon	Boiling point in °C
CH_4	−162
C_2H_6	−89
C_3H_8	−42
C_4H_{10}	−1
C_5H_{12}	36
C_6H_{14}	69
C_7H_{16}	98
C_8H_{18}	126

11.1 Identify the homologous series that the hydrocarbons in **Table 5** belong to. **[1 mark]**

11.2 Give the formula of the hydrocarbon in this homologous series with 9 carbon atoms. **[1 mark]**

11.3 Name the hydrocarbon with the chemical formula C_3H_8. **[1 mark]**

> **Exam Tip**
> You are told that all of the hydrocarbons are from one homologous series. Therefore, you only need to look at the first formula to identify the series.

11.4 The student carries out complete combustion with C_4H_{10}. Write the balanced symbol equation with state symbols for the reaction. **[3 marks]**

11.5 Draw a dot and cross diagram for the hydrocarbon CH_4. **[2 marks]**

> **Exam Tip**
> This is a covalent compound so needs overlapping circles.

12 Propane gas is used for camping stoves. Its formula is C_3H_8.

12.1 A cylinder contains 6.00 kg of propane. Calculate the volume that this mass of propane would occupy at room temperature and pressure. Give your answer to three significant figures. Relative atomic masses A_r: C = 12; H = 1 **[4 marks]**

12.2 Propane burns in air to make carbon dioxide and water:
$$C_3H_8 + 5O_2 \rightarrow 3CO_2 + 4H_2O$$
Deduce the volume of oxygen that reacts with 50 cm³ of propane. Give your answer in dm³. **[2 marks]**

12.3 480 g of propane is burnt in air. Calculate the volume of carbon dioxide produced at room temperature and pressure. Give your answer to three significant figures. **[4 marks]**

> **Exam Tip**
> If you do not give your answer to three significant figures, you will not get full marks.

13 A solution of glucose has a concentration of 55.6 mol/dm³. The formula of glucose is $C_6H_{12}O_6$.

13.1 Calculate the number of moles of glucose in 5.00 dm³ of the solution. **[1 mark]**

13.2 Deduce the concentration of the solution in kg/dm³. Relative atomic masses A_r: C = 12; O = 16; H = 1 **[4 marks]**

Exam Tip: Start by writing down the equation you are going to use.

13.3 Calculate the mass of glucose in 50.0 cm³ of solution. **[2 marks]**

14 Alkanes are a group of hydrocarbons.

14.1 The general formula for an alkane is C_nH_{2n+2}. Use the general formula to write the chemical formula of an alkane which contains five carbons. **[1 mark]**

14.2 Name the alkane with the chemical formula C_2H_6. **[1 mark]**

14.3 Draw the displayed formula of methane. **[2 marks]**

15 Crude oil is a finite resource found in rocks.

15.1 What is crude oil made of? Choose **one** answer. **[1 mark]**

a mixture of atoms

a pure solution of organic compounds

a mixture of different length hydrocarbons

a pure solution of a single length hydrocarbon

15.2 When crude oil is separated by distillation the resulting groups are called fractions. What does each fraction of crude oil contain? Choose **one** answer. **[1 mark]**

a single hydrocarbon

identical boiling points

similar length hydrocarbon chains

similar number of oxygen atoms

15.3 What property is used to separate the fractions of crude oil? Choose **one** answer. **[1 mark]**

boiling point   flammability

melting point   viscosity

15.4 Give **three** uses of the products of crude oil. **[3 marks]**

Exam Tip: 15.4 does not need lots of writing – a simple list of three things will get the marks.

15.5 Petrol, diesel, and residue are three fractions of crude oil.

Describe and explain how petrol, diesel, and residue are separated by fractional distillation. **[6 marks]**

C12

16 Table 6 shows the name and molecular formula of four alkanes.

Table 6

Name	Molecular formula
methane	CH_4
decane	$C_{10}H_{22}$
pentadecane	$C_{15}H_{32}$
icosane	$C_{20}H_{42}$

16.1 Use **Table 6** and the words in box to complete the sentences. **[5 marks]**

| flammable   highest   length   longest   lowest   shortest   viscous |

Some properties of hydrocarbons change depending on the

_____ of the hydrocarbon chain.

Methane has the _____ boiling point because it has

the _____ hydrocarbon chain.

Icosane is the most _____ because it has the

_____ hydrocarbon chain.

> **Exam Tip**
> Remember to use the molecular formulae in **Table 6** to help you answer **16.1**.

16.2 Which alkane in **Table 6** is the most flammable? Choose **one** answer. **[1 mark]**

decane   icosane   methane   pentadecane

16.3 Draw **one** line from each alkane to its correct melting point. **[3 marks]**

Alkane	Melting point
decane	−182 °C
icosane	−30 °C
methane	17 °C
pentadecane	36 °C

16.4 The complete combustion of an alkane gives two products. Complete the word equation for the complete combustion of methane. **[2 marks]**

methane + _____ → _____ + _____

> **Exam Tip**
> Unless the question asks for it, don't give the chemical symbols. If you make a mistake writing down a chemical symbol you won't get the marks.

Knowledge

C13 Other hydrocarbon products

Cracking

Not all hydrocarbons are as useful as each other. Longer molecules tend to be less useful than shorter ones. As such, there is a higher demand for shorter-chain hydrocarbons than longer-chain hydrocarbons.

A process called **cracking** is used to break up longer hydrocarbons and turn them into shorter ones.

Cracking produces shorter alkanes and **alkenes**.

Two methods of cracking are:

- Catalytic cracking – vaporise the hydrocarbons, then pass them over a hot catalyst.
- Steam cracking – mix the hydrocarbons with steam at a very high temperature.

Alkenes

Alkenes are a family of hydrocarbons that contain double bonds between carbon atoms.

They are called **unsaturated** and atoms can be added to the molecule by breaking the double bond. Alkanes are called **saturated** as there is no space to add more atoms.

The general formula for alkenes is C_nH_{2n}.

Alkenes are used as fuels, and to produce polymers and many other materials.

ethene, C_2H_4 propene, C_3H_6

They are much more reactive than alkanes. When mixed with bromine water, the **bromine water** turns from orange to **colourless**. This can be used to tell the difference between alkanes and alkenes.

Combustion reactions of alkenes

- Complete combustion produces carbon dioxide and water.
- Incomplete combustion is more likely, resulting in a smoky yellow flame.
- Both types of alkene combustion release less energy per mole than alkanes.

Polymers

Polymers are very long molecules made up of lots of smaller molecules joined together in a repeating pattern. The smaller molecules are called monomers. The process of turning many monomers into a polymer is called polymerisation.

Addition polymerisation

Addition polymerisation starts with molecules with a C=C bond (e.g., alkenes) as the monomer. The carbon-carbon double bond breaks in each molecule, and the carbon atoms then link together.

many single ethene monomers → long chain of poly(ethene) where n is a large number

The n refers to a large number of molecules. The rounded brackets and the bonds sticking out of them represent where the next molecule in the chain goes.

The inside of the brackets is known as the repeating unit – the section that repeats over and over again many thousands of times in the polymer.

Addition polymers are named after the monomer used to create them.
- An addition polymer made of ethene is called poly(ethene).
- An addition polymer made of propene is called poly(propene).

Key Terms

Make sure you can write a definition for these key terms.

addition   alkene   biodegradable   bromine water   cracking   colourless
polymerisation   saturated   thermosetting   thermosoftening   unsaturated

C13

Properties of polymers

The properties of polymers depend on:
- the monomers that make them up
- the conditions under which they are made.

For example, low-density poly(ethene) and high-density poly(ethene) are both made from ethene monomers but have very different properties due to the way that the polymer chains line up in the material.

Low-density poly(ethene)

LDPE is formed when the addition polymerisation reaction of ethene is carried out under high pressure and in the presence of a small amount of oxygen.

The branched polymer chains cannot pack together, so causing the low density of the polymer.

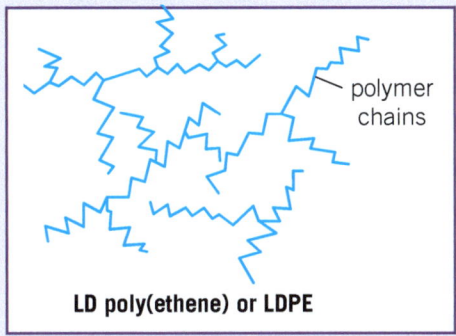

LD poly(ethene) or LDPE

High-density poly(ethene)

HDPE is formed when the addition polymerisation reaction of ethene is carried out using a catalyst at 50 °C. The polymer chains are straight and can pack tightly together, so causing the high density of the polymer.

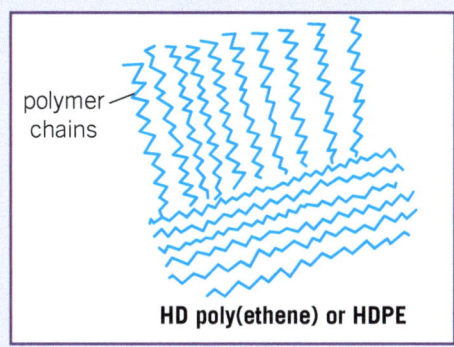

HD poly(ethene) or HDPE

Thermosoftening polymers

Thermosoftening polymers do not have links between the different chains, and soften when they are heated.

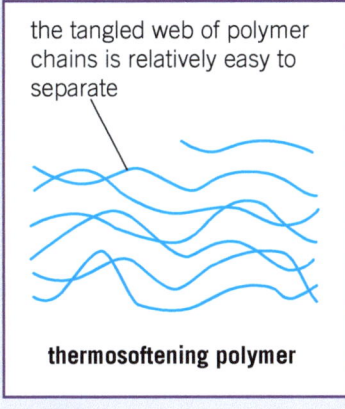

thermosoftening polymer

Thermosetting polymers

Thermosetting polymers have strong links between the different chains, and do not melt when they are heated.

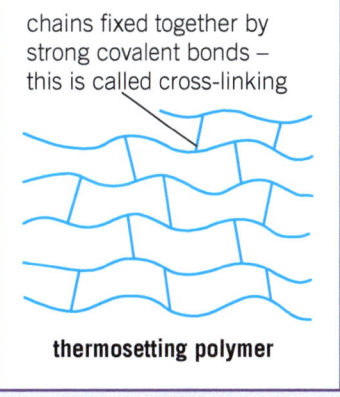

thermosetting polymer

Biodegradability of polymers

Many polymers are not **biodegradable**, so they are not broken down by microbes. This can lead to problems with waste disposal. Biodegradable plastics made from cornstarch have been developed. These will break down naturally over time as microbes can break the polymer chains up.

Uses of polymers

Polymers have many useful applications and new uses are being developed. Examples include:

- new packaging materials
- waterproof coatings for fabrics
- dental polymers
- wound dressings
- smart materials (including shape memory polymers)

C13 Knowledge

Retrieval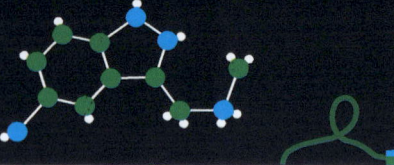

Learn the answers to the questions below then cover the answers column with a piece of paper and write as many as you can. Check and repeat.

C13 questions | Answers

#	Question	Answer
1	What is cracking?	breaking down a hydrocarbon with a long chain into smaller molecules
2	Why is cracking important?	it can produce more useful substances than long chain hydrocarbons
3	Name two methods to carry out cracking.	steam cracking and catalytic cracking
4	How is steam cracking carried out?	hydrocarbon is mixed with steam at high temperature
5	How is catalytic cracking carried out?	hydrocarbon is vapourised and passed over a hot catalyst
6	What are the products of cracking?	short chain alkanes and alkenes
7	Why are short chain alkanes more useful than long chain alkanes?	they are in more demand for use as fuels
8	What is the positive test for an alkene?	add bromine water: colour change from brown to colourless
9	What is a polymer?	a long-chain molecule made from repeating monomers
10	Why can thermosoftening plastics melt when heated?	thermosoftening plastics have weak intermolecular forces between the polymer chain, which are easy to break
11	Why do thermosetting plastics not melt when heated?	thermosetting plastics have strong links between the polymer chains, which are hard to break
12	How do you make high-density poly(ethene)?	50 °C and a catalyst
13	What does biodegradable mean?	can be broken down naturally by microbes
14	Give four new uses of polymers.	new packaging materials, waterproof coatings for fabrics, dental polymers, wound dressings, smart materials (including shape memory polymers)

C13 Other products from fuels

C13

Now go back and use the questions below to check your knowledge from previous chapters.

Previous questions | Answers

#	Question	Answer
1	What is a catalyst?	a substance that increases the rate of a reaction but is not used up in the reaction
2	How do catalysts increase the rate of a reaction?	they lower the activation energy of the reaction, so more collisions result in a reaction
3	Why is the rate of a reaction highest at the start of a reaction?	there are lots of reactant particles so frequent collisions
4	What is meant by chain length?	how many carbon atoms there are in an alkane
5	How does boiling point depend on chain length?	longer the chain, higher the boiling point
6	How does flammability depend on chain length?	longer the chain, lower the flammability
7	How can the different alkanes in crude oil be separated?	fractional distillation
8	In terms of energy transfer, what is an endothermic reaction?	one which transfers energy from the surroundings
9	What happens to the temperature of the surroundings in an exothermic reaction?	increases
10	Give an example of an exothermic reaction.	combustion, oxidation, neutralisation
11	What is the test for hydrogen?	squeaky pop
12	What is the test for oxygen?	relights a glowing splint
13	What is the test for carbon dioxide?	turns limewater milky if bubbled through it
14	What is the test for chlorine?	bleaches damp litmus paper

(Put paper here)

Required Practical Skills

Practise answering questions on the required practicals using the example below. You need to be able to apply your skills and knowledge to other practicals too.

Testing for alkenes

The standard laboratory test for an alkene is the addition of bromine water.

You need to be able to describe how to carry out the test, what observation you would make, and what conclusion you can draw from it.

Remember: even a negative result needs a conclusion to be drawn from it.

 Revision Tip

Make sure you always describe alkenes as 'colourless'. Never use the words 'white' or 'clear'.

Worked example

To perform this test, a few drops of bromine water are added to an unknown sample and shaken. If the orange colour disappears then the sample contains an alkene.

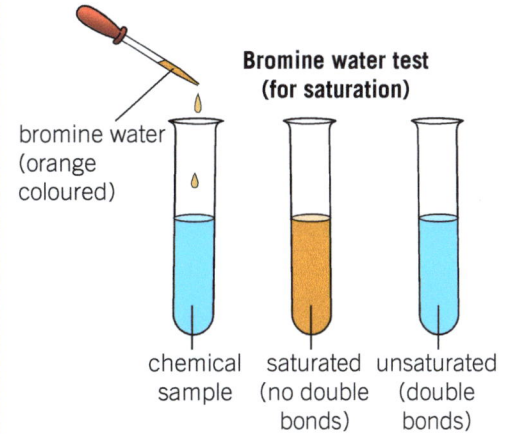

Practice

1. State the test for an alkene.
2. A student has three unmarked test tubes A, B, and C. They suspect one is an alkene. Describe a simple procedure they could follow to determine which sample was an alkene.

Practice

Exam-style questions

01 Some students use the apparatus shown in **Figure 1** to crack hydrocarbons.

Figure 1

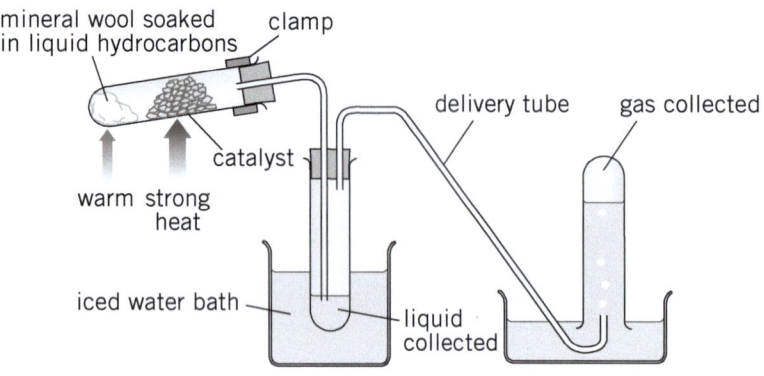

Exam Tip

Think about the changes that would happen if you just heated the liquid hydrocarbons.

01.1 Give a reason for heating the mineral wool soaked in liquid hydrocarbons. **[1 mark]**

_____

01.2 Identify the error in how the students set up the delivery tube. **[1 mark]**

_____

01.3 The equation shows one reaction that occurs in the apparatus.

$$C_{10}H_{22} \rightarrow C_5H_{12} + C_3H_6 + C_2H_4$$

One of the products of the reaction above is collected as a liquid. Write down the formula of the product that you predict is collected as a liquid. Give a reason for your prediction. **[2 marks]**

Exam Tip

Look at the length of the hydrocarbons to help your prediction.

_____

01.4 The product with the formula C_3H_6 is an alkene. Describe the colour change that occurs when C_3H_6 reacts with bromine water. **[1 mark]**

_____

02 A chemist does some tests on four hydrocarbons. The hydrocarbons have the formulae below. The chemist does not know which hydrocarbon is which.

C_2H_4 C_2H_6 C_8H_{18} $C_{17}H_{36}$

Table 1 shows the results of the tests.

Table 1

Hydrocarbon	Boiling point in °C	Observations when shaken with bromine water
A	126	no change
B	−104	orange to colourless
C	302	no change
D	−89	no change

152 C13 Other products from fuels

C13

02.1 Deduce the formula of each hydrocarbon in **Table 1**. Justify your decision. **[4 marks]**

> **Exam Tip**
> Use the general formula for alkanes and alkenes to first work out which compound belongs to which homologous group.

02.2 Predict the letter of the hydrocarbon in **Table 1** that is most viscous in the liquid state. **[1 mark]**

02.3 In a cracking reaction, a molecule of $C_{20}H_{42}$ forms two compounds:

C_8H_{18} C_3H_6

Write a balanced chemical equation for the cracking reaction. Do **not** include state symbols. **[3 marks]**

> **Exam Tip**
> Start by trying to find the number of each product by looking at the carbon atoms.

03 This question is about cracking.

03.1 Compare the conditions used for steam cracking and for catalytic cracking. **[3 marks]**

03.2 The equation shows a cracking reaction:

$C_{10}H_{22} \rightarrow C_6H_{14} + C_2H_4$

Balance the equation by writing a number where required. **[1 mark]**

> **Exam Tip**
> Start by balancing the carbons; the hydrogens should fall in line behind that.

03.3 Give **two** reasons for carrying out cracking reactions in industry. **[2 marks]**

04 A student bubbles compound **A** through bromine solution.

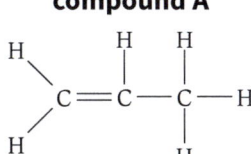

compound **A**

04.1 Describe the observation that the student would make. **[1 mark]**

> **Exam Tip**
> For **04.1** you need to describe a colour change. It is not enough to say the solution turns clear. If you don't use the correct terms, you will not get the mark.

04.2 Draw the structural formula of the product formed in the reaction of compound **A** with bromine solution. **[1 mark]**

04.3 A new sample of compound **A** is reacted with hydrogen gas. Describe the conditions required for the reaction. **[2 marks]**

04.4 Draw the structural formula of the product formed in the reaction of compound **A** with hydrogen and give its name. **[2 marks]**

C13 Practice 153

05 This question is about alkanes and alkenes.

05.1 Name the alkane with the formula C_2H_6. **[1 mark]**

05.2 Name the compound with the formula shown in **Figure 2**. **[1 mark]**

Figure 2

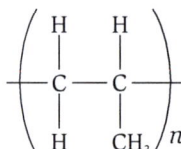

05.3 Deduce the formula of the alkene that has 8 carbon atoms. **[1 mark]**

05.4 Deduce the formula of the alkane that has 12 hydrogen atoms. **[1 mark]**

06.1 Polymers are useful substances used in everyday life. Define polymer. **[2 marks]**

06.2 **Figure 3** shows the repeating unit of a polymer.

Figure 3

Identify which monomer is needed to make the polymer in **Figure 3**. Choose **one** answer. **[1 mark]**

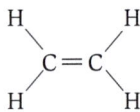 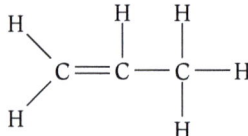

> **Exam Tip**
> On **Figure 3**, draw another line between the two carbons, then scribble out the lines touching the brackets – now try to match it to one of the answer options.

> **Exam Tip**
> There will be more than one line going into each box on the right-hand side.

06.3 Give the name of the polymer formed from the repeating unit shown in **Figure 4**. **[1 mark]**

Figure 4

> **Exam Tip**
> First identify the monomer. This will help you to name the polymer.

07 **Figure 5** shows a section of a polymer. The polymer was made by addition polymerisation.

Figure 5

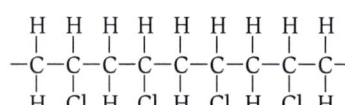

07.1 Draw the repeating unit of the polymer. [1 mark]

07.2 Draw the formula of the monomer used to make the polymer. [1 mark]

> **Exam Tip**
> Draw circles on **Figure 5** to identify the repeating unit, then redraw that with a double bond in the middle.

07.3 Polymers like those in Figure 5 are not biodegradable. Define *biodegradable* and explain why this is a problem. [2 marks]

08 PTFE is a polymer made form the monomer shown in **Figure 6**.

Figure 6

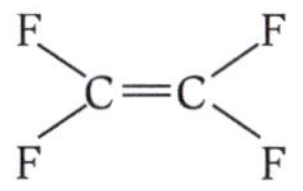

08.1 Identify the feature in **Figure 6** that means it can undergo addition polymerisation. [1 mark]

08.2 Draw a section of the polymer made from four monomer molecules. [1 mark]

08.3 The name of the monomer is tetrafluoroethene. Deduce the name of the polymer it forms. [1 mark]

> **Exam Tip**
> Don't be put off by the fluorines instead of hydrogens; it acts in exactly the same way. You just need to apply what you know to a new context.

09 Companies crack long-chain hydrocarbons to produce more useful shorter-chain hydrocarbons. The equation shows an example of a cracking reaction.

$C_{19}H\_\_\_ \rightarrow C_8H_{20} + C\_\_\_H_{10} + C_6H_{10}$

09.1 Complete the chemical formulae in the symbol equation. [1 mark]

09.2 Name the homologous series that the original compound belongs to. [1 mark]

09.3 Identify the product from the cracking reaction that is an alkane. [1 mark]

09.4 Identify the product from the cracking reaction that has two double bonds. [1 mark]

09.5 Complete the diagram to show the possible structure for the compound with two double bonds. [3 marks]

C—C—C—C—C

> **Exam Tip**
> Each carbon will only make four bonds in total, double bonds count as 2. Each hydrogen will only ever make one bond.

09.6 Describe the conditions required for catalytic cracking. [1 marks]

10 This question is about ethane, C_2H_6, and ethene, C_2H_4.

10.1 Draw a dot and cross diagram to show the bonds in ethene.
[3 marks]

10.2 Table 2 gives the energy needed to break single and double carbon–carbon bonds. Give a reason for the greater strength of the carbon–carbon double bond. **[1 mark]**

Table 2

Bond	Bond energy in kJ/mol
C—C	348
C=C	642

10.3 Suggest why ethene takes part in more reactions than ethane, even though ethene has the stronger carbon–carbon bond. **[2 marks]**

11 PVA glue contains the polymer poly(vinyl alcohol). **Figure 7** shows part of the structure of poly(vinyl alcohol).

Figure 7

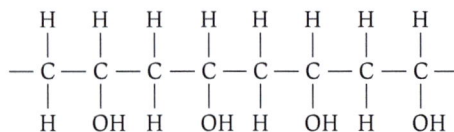

11.1 Draw the monomer of poly(vinyl alcohol). **[1 mark]**

11.2 Identify the type of polymerisation that forms poly(vinyl alcohol). **[1 mark]**

> **Exam Tip**
> Draw a circle around what you think is the repeating unit. Remember, the monomer needs to have a functional group to react to form the polymer.

11.3 PVA glue is a thick liquid. When the chemical borax is added to PVA glue, the polymer shown in **Figure 8** is formed.

Figure 8

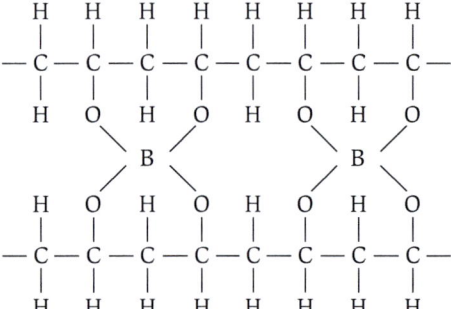

Predict how the properties of the polymer in **Figure 8** would differ from poly(vinyl alcohol). Explain your answer. **[3 marks]**

12 Both low-density poly(ethene) (LDPE) and high-density poly(ethene) (HDPE) are made from the same starting materials.

12.1 Draw the repeating unit of poly(ethene). **[2 marks]**

12.2 Draw the monomer that forms poly(ethene). **[1 mark]**

12.3 Compare the structures of LDPE and HDPE. **[4 marks]**

12.4 Both LDPE and HDPE are thermosoftening polymers. Explain why LDPE and HDPE are thermosoftening polymers. **[2 marks]**

> **Exam Tip**
> Remember, the monomer will be the substance that's within the brackets of the polymer name. In this instance, it's ethene.

C13 Other products from fuels

13 Lead is found naturally as lead sulfide, PbS. Lead sulfide is mixed with other substances in rock. This rock is called lead ore. Lead is extracted from lead ore by the steps below.

1. Lead sulfide is separated from the substances it is mixed with in lead ore.
2. Pure lead sulfide is heated with oxygen. This chemical reaction makes lead oxide and sulfur dioxide.
3. The lead oxide from step **2** is heated with carbon.

13.1 A lead ore contains 25% lead sulfide.

Calculate the mass of lead in 240 kg of this ore. Give your answer to two significant figures. **[4 marks]**

> **Exam Tip**
> First determine the mass of lead sulfide within the rock, then determine the percentage by mass of lead within lead sulfide – combine these to get the final answer.

13.2 Write a balanced symbol equation for step **2**. Include state symbols in your equation. **[3 marks]**

13.3 Explain whether lead oxide is oxidised or reduced in step **3**. **[2 marks]**

13.4 Explain why lead is extracted from its oxide by heating with carbon but aluminium cannot be extracted from its oxide in this way. **[2 marks]**

13.5 Name **one** other metal that can be extracted from its oxide by heating with carbon. **[1 mark]**

> **Exam Tip**
> Remember OILRIG:
> **O**xidation
> **I**s
> **L**oss (of electrons)
> **R**eduction
> **I**s
> **G**ain (of electrons)

14 A student wants to find the position of nickel in the reactivity series. The student adds small pieces of iron, lead, and nickel to dilute hydrochloric acid and to water. The student's observations are shown in **Table 3**.

Table 3

Metal	Observations on adding the metal to dilute hydrochloric acid	Observations on adding the metal to water and leaving for a few days
iron	bubbles form slowly on the surface of the metal	red-brown flakes form on the surface of the metal
lead	no change	no change
nickel	bubbles form slowly on the surface of the metal	no change

14.1 Use the observations in **Table 3** to deduce the position of nickel in the reactivity series. Justify your decision. **[3 marks]**

> **Exam Tip**
> This is a common type of question in the exam. Practise it here.

14.2 The student wants to confirm the position of nickel in the reactivity series relative to iron.

Suggest how you could improve the experiment to confirm the position of nickel by using displacement reactions. In your answer, describe and explain the results you would expect. **[4 marks]**

14.3 Nickel will displace copper from a solution of copper(II) sulfate.

Write an ionic equation for the displacement reaction between nickel and copper sulfate. Identify which species is reduced. **[3 marks]**

> **Exam Tip**
> As well as balancing the elements, the charges need to be balanced as well.

Knowledge

C14 Alcohols, carboxylic acids, and esters

Homologous series

There are lots of different 'families' of carbon-containing compounds, for example, alkanes and **alkenes**. These families are called **homologous series**. Each compound within a homologous series has similar properties and reactions.

Functional groups

Each homologous series contains specific atoms in specific orders, called **functional groups**.

The **alcohol** functional group is: —OH.

The first three alcohols in the homologous series are: methanol, ethanol, propanol

The **carboxylic acid** functional group is:

The first three carboxylic acids in the homologous series are: methanoic acid, ethanoic acid, propanoic acid

Formation of ethanol

Ethanol can be formed from the **fermentation** of sugar – warm a sealed mixture of yeast and a sugar solution.

glucose → ethanol + carbon dioxide

$C_6H_{12}O_6(aq) \rightarrow 2C_2H_5OH(aq) + 2CO_2(g)$

Uses of alcohols

- The first four alcohols in the homologous series mix easily with water to make a neutral solution, so are used as solvents for substances that don't dissolve in water.
- Common in perfumes, aftershaves, and mouthwashes
- Biofuels are a renewable alternative to fossil fuels, which have a reduced carbon footprint. They do however divert land use from food production.

Combustion reaction

- Short alcohols are very effective fuels and combust easily, burning with a blue flame and producing carbon dioxide and water:

$$2CH_3OH + 3O_2 \rightarrow 2CO_2 + 4H_2O$$

- Carboxylic acids can also undergo combustion, but we do not generally do this or use them as a fuel.

C14

Formation and uses of carboxylic acids

Carboxylic acids are formed by the oxidation of alcohols. For example, ethanol can be **oxidised** to ethanoic acid (a carboxylic acid), either by chemical **oxidising agents** or by microbial action.

$$\underset{\text{ethanol}}{H-\underset{\underset{H}{|}}{\overset{\overset{H}{|}}{C}}-\underset{\underset{H}{|}}{\overset{\overset{H}{|}}{C}}-OH} \xrightarrow{[O]} \underset{\text{ethanoic acid}}{H-\underset{\underset{H}{|}}{\overset{\overset{H}{|}}{C}}-\overset{\overset{O}{\|}}{C}-OH}$$

Ethanoic acid is used in vinegar for cooking and cleaning. Carboxylic acids can also be used to make salts and esters.

Weak acids

When added to water, carboxylic acids are partially ionised to form weakly acidic solutions.
They are weak acids.
An aqueous solution of a weak acid has a higher pH than an aqueous solution of a strong acid with the same concentration.

Reaction with alcohols

Carboxylic acids react with alcohols to make water and **esters**. The reaction requires sulfuric acid as a catalyst.

Esters have the functional group —COO— and their names end with '—oate'. They have distinctive smells and are used in perfumes and flavourings. The product of ethanol and ethanoic acid is ethyl ethanoate.

Reaction with sodium

Alcohols react with sodium to release hydrogen. The product from this reaction is called an **alkoxide**, which if added to water forms a strongly alkaline solution.

Reaction with sodium carbonate

Carboxylic acids react with bases to form salts. For example, carboxylic acids react with a metal carbonate to produce a salt, carbon dioxide, and water.

 **Key Terms**

Make sure you can write a definition for these key terms.

alcohols   alkene   alkoxide   carboxylic acid   ester   fermentation
functional group   homologous series   oxidation   oxidising agent

Retrieval

Learn the answers to the questions below then cover the answers column with a piece of paper and write as many as you can. Check and repeat.

C14 questions | Answers

#	Question	Answer
1	What is a homologous series?	a group of compounds with the same functional group
2	What is a functional group?	a group of atoms that determines the properties of a compound
3	What are alcohols?	a homologous series with a carbon-oxygen-hydrogen bond
4	How are alcohols produced?	reaction of steam with an alkene, or fermentation
5	What conditions are required to produce alcohols by fermenting?	sugar solution with yeast mixed in, warm, sealed vessel
6	Name the first three alcohols.	methanol, ethanol, propanol
7	What are the products of a reaction between an alcohol and sodium?	hydrogen and an alkoxide
8	What can be observed when sodium is added to an alcohol?	fizzing
9	What is the oxidation of an alcohol?	when an alcohol is turned into a carboxylic acid
10	Name two oxidising agents.	acidified potassium dichromate, microbial agents
11	What is a carboxylic acid?	a homologous series with a —COOH group
12	How are carboxylic acids produced?	oxidation of alcohols
13	Name the first three carboxylic acids.	methanoic acid, ethanoic acid, propanoic acid
14	How can you test for carboxylic acids?	react with sodium carbonate, fizzes (the gas is carbon dioxide)
15	What is the product of a reaction between a carboxylic acid and an alcohol?	an ester
16	What catalyst is normally used in the formation of esters?	concentrated sulfuric acid
17	What are esters used for?	perfumes and flavourings
18	What occurs when pure carboxylic acids are added to water?	a weak acid is formed

C14 Alcohols, carboxylic acids, and esters

C14

Now use the questions below to check your knowledge from previous chapters.

Previous questions | Answers

#	Question	Answer
1	What is the relative mass of a proton?	1
2	In an endothermic reaction, will the overall energy change be positive or negative?	positive
3	What is a chemical cell?	a device that uses chemical reactions to generate electricity
4	How are chemical cells made?	by connecting two different metals (electrodes) in a solution (electrolyte)
5	What is a battery?	two or more chemical cells connected in series
6	What is a hydrocarbon?	a compound containing carbon and hydrogen only
7	In terms of the substances that make it up, what is crude oil?	a mixture of different compounds, mostly hydrocarbons
8	What is crude oil used for?	fuels and as feedstock for other industries
9	What is a reversible reaction?	one where the reactants turn into products and the products turn into reactants
10	What is a closed system?	a reaction where no products or reactants are allowed to escape
11	What is dynamic equilibrium?	the point in a reversible reaction when the rate of the forward and reverse reactions are the same
12	What are the three external conditions that can be changed?	concentration, temperature, pressure
13	What is the effect of increasing the concentration of reactants on a reaction at dynamic equilibrium?	favours the forward direction

Maths Skills

Practise your maths skills using the worked example and practice questions below.

Finding the mean

Whenever an experiment is conducted, it is important to repeat it to establish how *precise* the values are (how close to each other they are) and how *repeatable* they are (can they be repeated).

Whenever you repeat an experiment and record repeat observations you must calculate a mean to give an average result for that observation. However, only use values that are close together, and discard any anomalous values.

Worked example

A student burns ethanol and uses the heat released to warm up some water. As soon as the water increases by 10 °C, the student stops the reaction and measures the mass of ethanol used. They repeat this three more times and record the masses: 5.1 g, 6.3 g, 6.5 g, 6.2 g.

Calculate the mean of the values.

Step 1: Establish which values to use – in this case 6.3, 6.5, and 6.2 are used, but 5.1 is ignored because it is not close to the others.

Step 2: Calculate the mean.

$$\text{mean} = \frac{\text{sum of values}}{\text{total number of values}}$$

$$= \frac{(6.3 + 6.5 + 6.2)}{3} = 6.3\,g$$

Practice

1. A student measures how the mass of a magnesium strip increases when burnt in oxygen. They record the masses: 0.12 g, 0.12 g, 0.14 g, 0.11 g, 0.23 g. Calculate the mean increase in mass.

2. The volume of gas produced in three repeats of an experiment is collected, and recorded as: 54 cm³, 58 cm³, 55 cm³. Calculate the mean volume of gas produced.

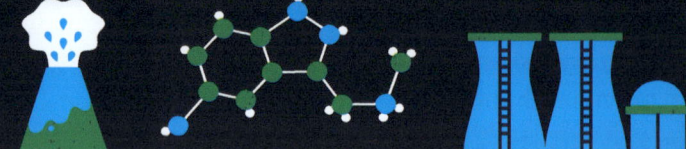

Exam-style questions

01 Figure 1 shows the boiling points of some alkanes and alcohols.

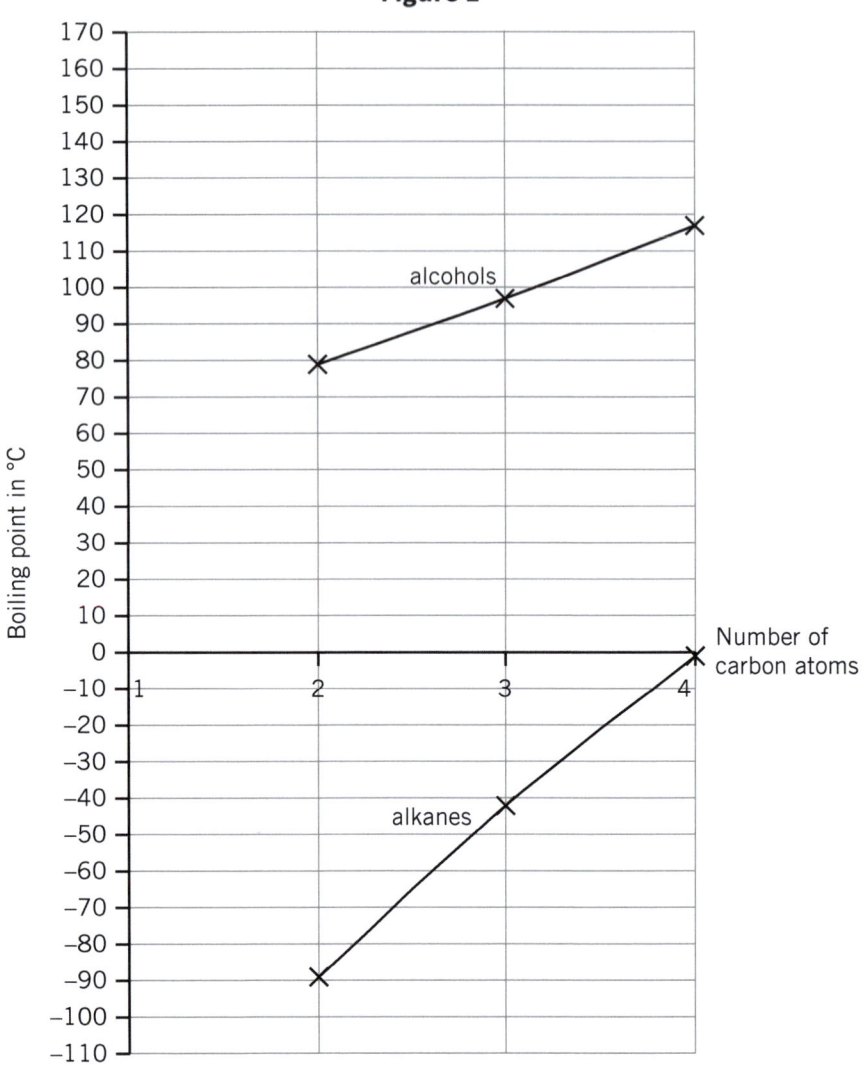

Figure 1

01.1 Name the compound with a boiling point of −42 °C. **[1 mark]**

_____

01.2 Name the compound with a boiling point of 79 °C. **[1 mark]**

_____

01.3 Table 1 shows the boiling points of some carboxylic acids.

Table 1

Number of carbon atoms	Boiling point in °C
2	118
3	141
4	164

Plot the boiling points of the carboxylic acids on **Figure 1**. Draw a line of best fit. **[2 marks]**

162   C14 Alcohols, carboxylic acids, and esters

C14

01.4 Describe **two** patterns shown in **Figure 1**. [2 marks]

1 _____

_____

2 _____

_____

01.5 Explain the pattern shown in the boiling points for the three alcohols. [2 marks]

_____

_____

_____

02.1 Draw **one** line from the name of each compound to its displayed formula. [4 marks]

Name	Displayed formula
ethene	H₂C=CH₂
ethanol	H-C-C-C-H (propane structure)
propane	H-C-C-O-H (ethanol structure)
propanol	H-C-C-C-O-H (propanol structure)
	H-C-C=C-H (propene structure)

! **Exam Tip**

The are two parts to a hydrocarbon name.
The first part is the number of carbon atoms:
- meth-   1 carbon
- eth-   2 carbons
- prop-   3 carbons

The seconds part that tells us the functional groups:
- -ene   has double bonds
- -ane   has single bonds
- -ol   is an alcohol with an –OH group

C14 Practice 163

02.2 Which homologous series does the compound with the formula CH₃CH₂COOH belong to?

Tick **one** box. [1 mark]

alcohols ☐   alkenes ☐

alkanes ☐   carboxylic acids ☐

02.3 Write the name of the compound with the formula C₄H₈. [1 mark]

_____

03 A student has a test tube of ethanoic acid. The student adds some sodium carbonate powder to the test tube.

03.1 Explain **one** observation the student would make. [2 marks]

03.2 A teacher mixes some ethanoic acid and ethanol together, then adds a few drops of a catalyst and warms the mixture. A reaction occurs and an ester is formed. Give the purpose of the catalyst. [1 mark]

03.3 Name the ester formed in the reaction. [1 mark]

03.4 Describe how the ester made can be detected. [1 mark]

04 The hydrogen atoms in the methyl group of ethanoic acid can be replaced with chlorine atoms. **Table 2** gives the pH of some of these acids.

Table 2

Acid	Formula	pH of acid with concentration of 1 mol/dm³	Acid	Formula	pH of acid with concentration of 1 mol/dm³
A	H—C(H)(H)—C(=O)(O—H)	2.42	C	Cl—C(H)(Cl)—C(=O)(O—H)	0.65
B	H—C(H)(Cl)—C(=O)(O—H)	1.44	D	Cl—C(Cl)(Cl)—C(=O)(O—H)	0.32

04.1 Deduce the effect of the presence of chlorine atoms on the degree of ionisation of a carboxylic acid. Justify your answer. [3 marks]

04.2 A student adds drops of acids **A** and **D**, separately, onto lumps of calcium carbonate. Compare the predicted observations for the two acids. [2 marks]

04.3 When ethanoic acid reacts with calcium carbonate, the salt made is calcium ethanoate. Predict the name of the salt made when the acid with the formula CH₃CH₂COOH reacts with calcium carbonate. [1 mark]

> **Exam Tip**
> For both **02.2** and **02.3**, look for the functional group, then count up the number of carbon atoms.

> **Exam Tip**
> Ethanoic acid will react in the same way as any other acid.

> **Exam Tip**
> pH is a value showing how many H⁺ ions have dissociated from the compound. The more H⁺ ions that dissociate the more ionised the carboxylic acid will be.

C14 Alcohols, carboxylic acids, and esters

C14

05 A student carried out some reactions of three organic compounds pentene, ethanoic acid, and ethanol. The compounds were labelled **X**, **Y**, and **Z**. The student did not know which compound was which. Their observations are shown in **Table 3**.

Table 3

Compound	Add sodium	Add sodium carbonate	Burn the compound
X	bubbles slowly	no reaction	burns with a blue flame that is hard to see
Y	no reaction	no reaction	burns with a smoky flame
Z	bubbles quickly	bubbles	

05.1 Deduce the identities of **X**, **Y**, and **Z**. Justify your decisions. **[6 marks]**

05.2 The student carries out another experiment. They boil some propanol with an oxidising agent. Name the homologous series that propanol is in. **[1 mark]**

05.3 Deduce the name of the organic product of the reaction of propanol with the oxidising agent. **[1 mark]**

> **! Exam Tip**
> Remember to use the evidence from **Table 3** to support your reasoning.

06 A student wanted to compare the energy transferred when different alcohols burn in air. They measured the temperature change of the water when different alcohols were burnt in a spirit burner. **Figure 2** shows the apparatus.

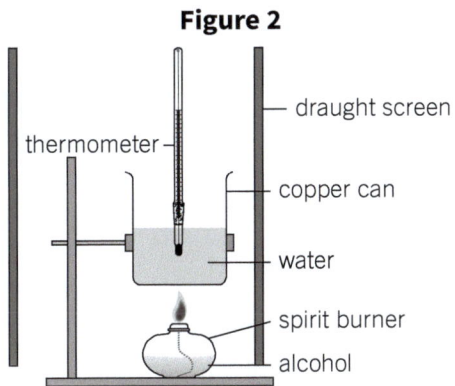

Figure 2

06.1 Identify the independent variable in the investigation. **[1 mark]**

06.2 Suggest **two** control variables in the investigation. **[2 marks]**

06.3 **Table 4** shows the student's results. Explain the trend shown by the results. **[3 marks]**

Table 4

Alcohol	Temperature increase of the water in °C
methanol	6.5
ethanol	9.5
propanol	13.0
butanol	16.5

C14 Practice

06.4 Write a balanced chemical equation for the complete combustion of butanol, C_4H_9OH. **[3 marks]**

06.5 Pentanol has five carbon atoms. Complete the structure of pentanol. **[1 mark]**

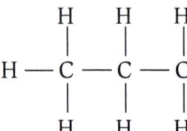

> **! Exam Tip**
>
> There are always the same two products from complete combustion. What will change is how the equation is balanced. Start with the carbons, then the hydrogens, leaving the oxygens to last.

07 Ethanol can be made in a fermentation reaction. The equation for the reaction is:

$$C_6H_{12}O_6 \rightarrow 2C_2H_5OH + 2CO_2$$

07.1 Name the living organism required for the fermentation reaction to occur. **[1 mark]**

07.2 Give the temperature at which fermentation is normally carried out. **[1 mark]**

07.3 Give **two** uses of ethanol. **[2 marks]**

07.4 In a fermentation reaction, 80.0 g of ethanol is made. Calculate the mass of glucose that reacted. **[5 marks]**

> **! Exam Tip**
>
> Carbon will always make four bonds, whilst hydrogens will always make one bond and oxygens will make two bonds.

> **! Exam Tip**
>
> The first step is to determine the M_r of glucose and ethanol.

08 Three compounds have the formulae **X**: $C_{10}H_{20}$, **Y**: $C_{10}H_{22}$, **Z**: $C_{10}H_{21}OH$

08.1 Calculate the relative formula mass of compound **Z**. Relative atomic masses A_r: C = 12; H = 1; O = 16 **[2 marks]**

08.2 Give the letter of the compound that reacts with steam to make an alcohol. **[1 mark]**

08.3 Write the formula of the product of the reaction of compound **X** with hydrogen. **[1 mark]**

08.4 Name the type of compound formed when compound **Z** reacts with ethanoic acid. **[1 mark]**

> **! Exam Tip**
>
> Use the general formulae to determine which homologous series the compounds belong to.

09 **Table 5** shows the solubility of some alcohols in water.

Table 5

Number of carbon atoms in alcohol	Name of alcohol	Solubility in g/per 100 g of water
1		the alcohol completely mixes with water
2	ethanol	
3	propanol	
4	butanol	8.14
5	pentanol	2.64
6	hexanol	0.592
7	heptanol	0.0928

09.1 Name the alcohol with one carbon atom. **[1 mark]**

09.2 Describe the pattern shown in **Table 5**. **[2 marks]**

09.3 Suggest a reason for the pattern shown in **Table 5**. **[1 mark]**

09.4 Calculate the solubility of pentanol in mol/dm³. Give your answer to two significant figures. Relative atomic masses A_r: C = 12; H = 1; O = 16 **[5 marks]**

10 This question is about the following compounds:
- **A** C_3H_6
- **B** C_3H_8
- **C** CH_3CH_2COOH
- **D** $CH_3CH_2CH_2OH$

10.1 Name compound **A**. **[1 mark]**

10.2 Name the homologous series that compound **D** belongs to. **[1 mark]**

10.3 Calculate the number of moles in 10.0 g of compound **C**. **[3 marks]**

10.4 Compare the chemical properties of compounds **C** and **D**. **[6 marks]**

> **! Exam Tip**
>
> Remember that the relative atomic mass of these elements are:
> C = 12
> H = 1
> O = 16

11 Ammonia, NH_3, is a gas at room temperature and pressure.

11.1 Give the number of moles of hydrogen atoms in one mole of ammonia gas. **[1 mark]**

11.2 Calculate the relative molecular mass of ammonia. **[2 marks]**

11.3 Calculate the number of moles of ammonia in a 68 g sample of the gas. **[2 marks]**

11.4 Calculate the number of particles of ammonia in a 68 g sample of the gas. The Avogadro constant is 6.02×10^{23} per mole. Give your answer to three significant figures. **[3 marks]**

> **! Exam Tip**
>
> You'll need to look up the mass numbers on the Periodic Table.

12 A student investigated the thermal decomposition of calcium carbonate, $CaCO_3$. They heated 40 g of calcium carbonate in a test tube.

12.1 Calculate the relative formula mass of calcium carbonate. Relative atomic masses A_r: C = 12; H = 1; O = 16; Ca = 40 **[2 marks]**

12.2 Calculate the number of moles of calcium carbonate that the student heated. **[2 marks]**

12.3 Complete the balanced symbol equation for the reaction. **[1 mark]**
$$CaCO_3(s) \rightarrow \underline{}(s) + CO_2(g)$$

12.4 The student measures the mass at the end of the reaction. The mass has decreased. Explain why the mass has decreased. **[2 marks]**

> **! Exam Tip**
>
> Take each of the elements in carbon dioxide away from calcium carbonate and see what you have left over.

C14 Practice

13 Compare the displacement reactions of halogens and metals.

In your answer, include ideas about oxidation and reduction and refer to the two equations below.

Reaction 1 $Cl_2(aq) + 2NaBr(aq) \rightarrow 2NaCl(aq) + Br_2(aq)$

Reaction 2 $Mg(s) + CuSO_4(aq) \rightarrow MgSO_4(aq) + Cu(s)$ **[6 marks]**

> **! Exam Tip**
> The question is asking *two* things about *two* equations. Follow this structure for your answer:
> In reaction 1 ... is oxidised
> In reaction 1 ... is reduced
> In reaction 2 ... is oxidised
> In reaction 2 ... is reduced
> Don't forget to compare the two reactions at the end.

14 Table 6 gives the properties of some elements and compounds. Each substance is represented by a letter.

Table 6

Substance	Melting point in °C	Boiling point in °C	Does it conduct electricity in the solid state?	Does it conduct electricity in the liquid state?
A	−182	−162	no	no
B	801	1465	no	yes
C	650	1110	yes	yes
D	1710	2230	no	no

14.1 Give the letter of the substance that could be methane, CH_4. **[1 mark]**

14.2 Complete the dot and cross diagram for methane shown in **Figure 3**. **[2 marks]**

Figure 3

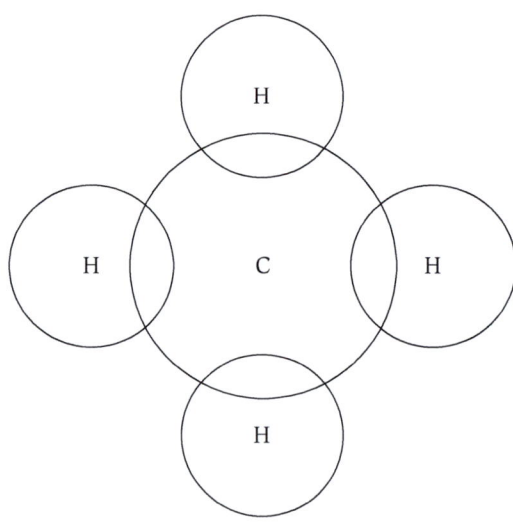

> **! Exam Tip**
> Start by adding an extra column to **Table 6** and filling it in with the type of substance: simple covalent, giant covalent, metallic, or ionic.

14.3 Explain which substance in **Table 6** could be sodium chloride. **[3 marks]**

14.4 Name and describe the structure and bonding in substance **C**. **[4 marks]**

C14 Alcohols, carboxylic acids, and esters

15 A student titrated 25.0 cm³ portions of dilute lithium hydroxide (LiOH) with a 0.210 mol/dm³ sulfuric acid solution. The equation for the reaction is:

$$2LiOH + H_2SO_4 \rightarrow Li_2SO_4 + 2H_2O$$

15.1 Which is the alkali in this reaction? Choose **one** answer. [1 mark]

LiOH   Li₂SO₄

H₂SO₄   H₂O

15.2 Name the salt formed in this reaction. [1 mark]

15.3 Give the ionic equation for this reaction. Include state symbols. [2 marks]

15.4 The student carries out four titrations. In each case, the student starts at 0.00 cm³. The final burette readings are shown in **Figure 4**. [2 marks]

Figure 4

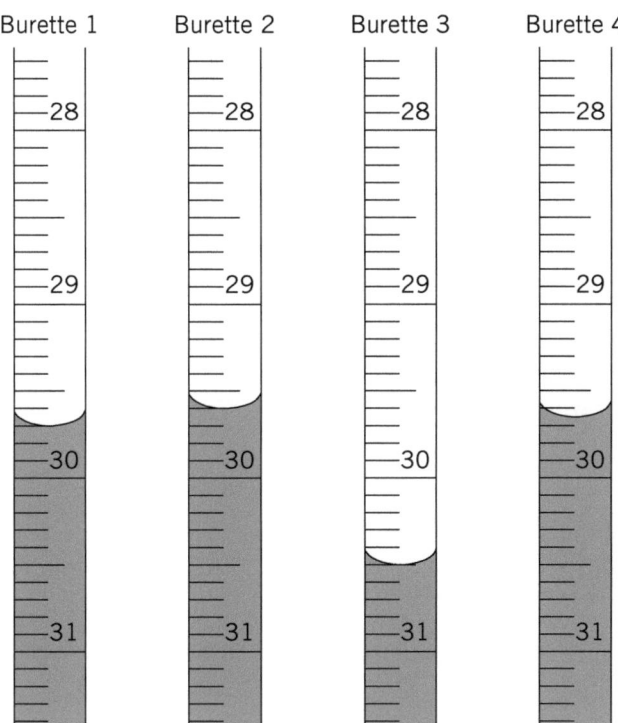

Complete **Table 7** by recording all the values to two decimal places to the nearest 0.05 cm³.

Table 7

	Titration 1	Titration 2	Titration 3	Titration 4
Volume of sulfuric acid in cm³				

15.5 Identify which value is not concordant (within 0.10 cm³ of each other) and calculate the mean volume of sulfuric acid used from the remaining values. [2 marks]

15.6 Give **one** reason why the values may not all be concordant. [1 mark]

15.7 The student titrated 25.0 cm³ portions of dilute lithium hydroxide (LiOH) with a 0.210 mol/dm³ sulfuric acid solution. Calculate the concentration of lithium hydroxide in mol/dm³ using the mean (from **15.5**). **[3 marks]**

16 This question is about energy changes in combustion reactions.

16.1 Methane will burn in a plentiful supply of oxygen to release energy. The equation for the reaction is:

$$CH_4 + 2O_2 \rightarrow CO_2 + 2H_2O$$

Draw a fully labelled reaction profile for the reaction on **Figure 4**.
[3 marks]

Figure 4

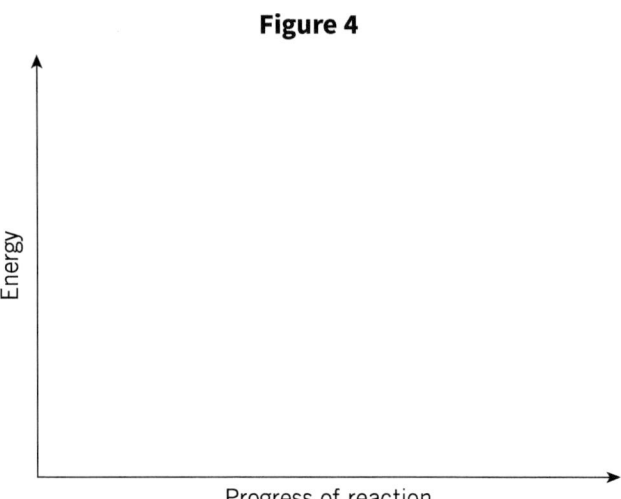

16.2 A student investigated the total energy released in combustion reactions of different chain-length alkanes. **Table 8** shows the student's results. The data for butane is missing.

Table 8

Name	Chemical formula	Number of carbons	Total energy released in kJ/mol
methane	CH_4	1	890
ethane	C_2H_6	2	1560
propane	C_3H_8	3	2220
butane	C_4H_{10}	4	
pentane	C_5H_{12}	5	3510
hexane	C_6H_{14}	6	4160

C14 Alcohols, carboxylic acids, and esters

Plot the data from **Table 8** on **Figure 5**. Include a line of best fit. **[3 marks]**

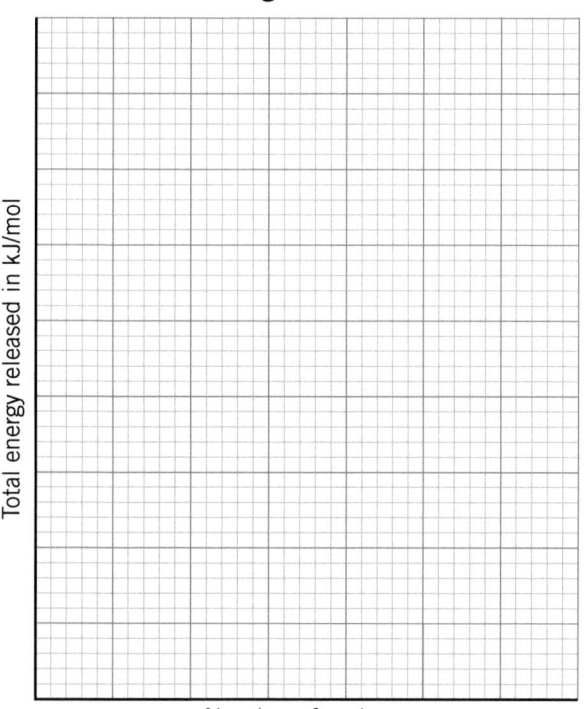

Figure 5

Total energy released in kJ/mol

Number of carbons

16.3 Use your graph to find the missing value for the energy released in the combustion of butane, C_4H_{10}. **[1 mark]**

16.4 **Figure 6** shows the displayed formulae equation for the combustion of methane.

Figure 6

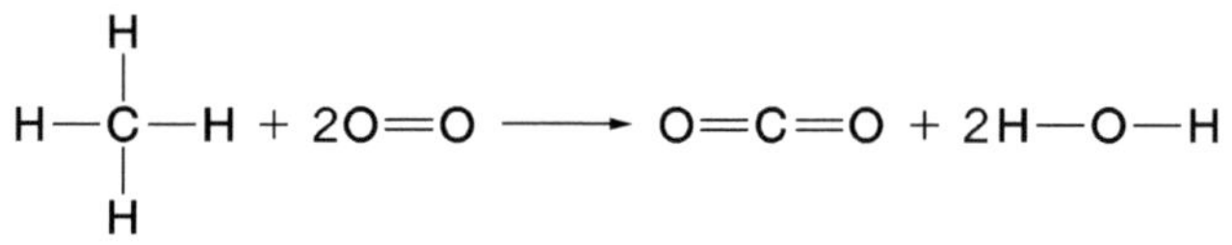

$$H-\underset{\underset{H}{|}}{\overset{\overset{H}{|}}{C}}-H + 2O=O \longrightarrow O=C=O + 2H-O-H$$

Table 9 shows the bond energies and the overall energy change in the reaction.

Table 9

	C—H	O=O	C=O	O—H	Overall energy change
Energy in kJ/mol	412	496	743	X	−890

Calculate the bond energy **X** for the O—H bond in kJ/mol.

Use **Figure 6** and **Table 9**. **[5 marks]**

16.5 The student stated that more energy is released when bonds are broken than is released when bonds are formed. Do you agree with the student? Explain your answer. **[2 marks]**

Periodic Table

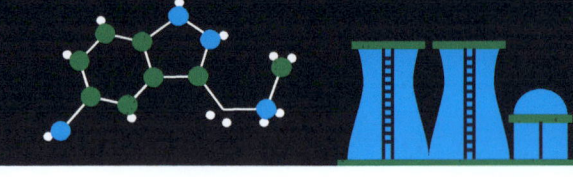

1	2											3	4	5	6	7	0
						1 **H** hydrogen 1											4 **He** helium 2
7 **Li** lithium 3	9 **Be** beryllium 4											11 **B** boron 5	12 **C** carbon 6	14 **N** nitrogen 7	16 **O** oxygen 8	19 **F** fluorine 9	20 **Ne** neon 10
23 **Na** sodium 11	24 **Mg** magnesium 12											27 **Al** aluminium 13	28 **Si** silicon 14	31 **P** phosphorus 15	32 **S** sulfur 16	35.5 **Cl** chlorine 17	40 **Ar** argon 18
39 **K** potassium 19	40 **Ca** calcium 20	45 **Sc** scandium 21	48 **Ti** titanium 22	51 **V** vanadium 23	52 **Cr** chromium 24	55 **Mn** manganese 25	56 **Fe** iron 26	59 **Co** cobalt 27	59 **Ni** nickel 28	63.5 **Cu** copper 29	65 **Zn** zinc 30	70 **Ga** gallium 31	73 **Ge** germanium 32	75 **As** arsenic 33	79 **Se** selenium 34	80 **Br** bromine 35	84 **Kr** krypton 36
85 **Rb** rubidium 37	88 **Sr** strontium 38	89 **Y** yttrium 39	91 **Zr** zirconium 40	93 **Nb** niobium 41	96 **Mo** molybdenum 42	[98] **Tc** technetium 43	101 **Ru** ruthenium 44	103 **Rh** rhodium 45	106 **Pd** palladium 46	108 **Ag** silver 47	112 **Cd** cadmium 48	115 **In** indium 49	119 **Sn** tin 50	122 **Sb** antimony 51	128 **Te** tellurium 52	127 **I** iodine 53	131 **Xe** xenon 54
133 **Cs** caesium 55	137 **Ba** barium 56	139 **La*** lanthanum 57	178 **Hf** hafnium 72	181 **Ta** tantalum 73	184 **W** tungsten 74	186 **Re** rhenium 75	190 **Os** osmium 76	192 **Ir** iridium 77	195 **Pt** platinum 78	197 **Au** gold 79	201 **Hg** mercury 80	204 **Tl** thallium 81	207 **Pb** lead 82	209 **Bi** bismuth 83	[209] **Po** polonium 84	[210] **At** astatine 85	[222] **Rn** radon 86
[223] **Fr** francium 87	[226] **Ra** radium 88	[227] **Ac*** actinium 89	[261] **Rf** rutherfordium 104	[262] **Db** dubnium 105	[266] **Sg** seaborgium 106	[264] **Bh** bohrium 107	[277] **Hs** hassium 108	[268] **Mt** meitnerium 109	[271] **Ds** darmstadtium 110	[272] **Rg** roentgenium 111	[285] **Cn** copernicium 112	[286] **Nh** nihonium 113	[289] **Fl** flerovium 114	[289] **Mc** moscovium 115	[293] **Lv** livermorium 116	[294] **Ts** tennessine 117	[294] **Og** oganesson 118

key

relative atomic mass
atomic symbol
name
atomic (proton) number

*The lanthanides (atomic numbers 58–71) and the actinides (atomic numbers 90–103) have been omitted.
Relative atomic masses for **Cu** and **Cl** have not been rounded to the nearest whole number.